DATE DUE

WOMEN PIONEERS OF PUBLIC EDUCATION

WOMEN PIONEERS OF PUBLIC EDUCATION

HOW CULTURE CAME TO THE WILD WEST

Jurgen Herbst

palgrave
macmillan

LA
230.5
.W48
H47
2008

Permission to publish Map of the Mountains between
Silverton and Durango on the book cover has been granted by
Drake Mountain Maps
18 Kaha Street
Rangataua, Ohakune
New Zealand

Permission to reprint Photographs 1.1, 2.1, 5.1, 8.1, 10.1, and 12.1
has been granted by
Fort Lewis College, Center of Southwest Studies,
Nina Heald Webber Southwest Colorado Collection.
Permission to use copy of report card granted by
San Juan County Historical Society.

First published in 2008 by
PALGRAVE MACMILLAN®
in the United States—a division of St. Martin's Press LLC,
175 Fifth Avenue, New York, NY 10010.

Where this book is distributed in the UK, Europe and the rest of the world,
this is by Palgrave Macmillan, a division of Macmillan Publishers Limited,
registered in England, company number 785998, of Houndmills,
Basingstoke, Hampshire RG21 6XS.

Palgrave Macmillan is the global academic imprint of the above companies
and has companies and representatives throughout the world.

Palgrave® and Macmillan® are registered trademarks in the United States,
the United Kingdom, Europe and other countries.

ISBN-13: 978–0–230–60835–1
ISBN-10: 0–230–60835–3

Library of Congress Cataloging-in-Publication Data is available from the
Library of Congress.

A catalogue record of the book is available from the British Library.

Design by Newgen Imaging Systems (P) Ltd., Chennai, India.

First edition: November 2008

10 9 8 7 6 5 4 3 2 1

Printed in the United States of America.

For Kris,
who loves the mountains as much as I do

CONTENTS

Figures

TABLES

INTRODUCTION

This is the story of a clash of cultures on the mining frontier of the San Juan Mountains in what is today the State of Colorado. The story took place during the thirty years from 1874 to 1903. It centered on the town of Silverton that came into being on the banks of the Animas River in a big, grassy meadow then called Baker's Park. It revolved around the fortunes of the town's public education from its first beginnings when all you had was a one-room school house, one teacher, two dozen students, and three school board members to supervise the enterprise. I follow the story to what was considered at the close of the century the most triumphal achievement of public education: the coming of the high school with its promise that every child whose parents so desired could in his hometown be prepared for college.

Prospectors and down-on-their-luck roughnecks from American and European mining areas had combed through the Rocky Mountain wilderness and in 1870 had discovered a lode of gold and silver ores in Arrastra Gulch, a creek flowing into the Animas River two miles north of Baker's Park. In the next two years they set up mining districts along the river from Eureka, six miles further north, to Baker's Park in the south. Once a quartz mill had begun operations in 1873 at the mine they called the Little Giant and a sawmill had been set up at Mineral Creek in Baker's Park, the town site of Silverton was surveyed and the first twenty-five cabins with their 100 inhabitants were erected.

A town at the site of a mining camp soon attracted technical and business entrepreneurs, both male and female, who were needed to set up a smelter, stores, hotels, a newspaper, a post office, blacksmith, barber, and other shops as well as saloons, dancing halls, and bordellos. They were joined by surveyors, engineers, lawyers, doctors, and other professionals and tradesmen. But most important for our story of a clash of cultures were the wives who soon agitated for schools for their children, for libraries and churches, and who began to establish literary and drama clubs to bring culture and civilization to the mountain wilderness in which they lived.

In Silverton it was the school house of the San Juan County Public School District No. 1 that became the center of the town's efforts to polish the rough edges of frontier and mining life. Public education was the civilizing force that was to tame the wilderness and steel the town's children against the temptations of "demon rum" and prostitution, evidences of which they saw around them daily in the town's dance halls and saloons.

When the school was established in 1876 there were then no compulsory education laws, no truant officers to corral the wayward youngsters and drive them into school. The town's wives of the owners of smelter and stores beat the drum for public education. They had the help of their husbands and a supporting press. They urged the town's mothers and fathers over and over again to send their children to school and to keep them there. Public education was the antidote to moral decay and lawlessness. It was the transforming power that would make the town's children civilized human beings and introduce them to mankind's heritage of literature, art, and music. It was the avenue over which they would travel to explore their country and all the world beyond their mountains.

This book, then, presents a chapter in the history of American education that differs from the usual accounts we have of the establishment of public schools by the male makers and shakers, the Horace Manns and Henry Barnards in New England, and the Calvin Stowes and John Pierces of the Midwest. Though here in the West men also established and served in public offices, it was women who in Silverton became the bearers of literary culture and civilized behavior. They saw to it that their and their fellow pioneers' children received an education that would enable them to endure and escape the barbarism of their natural and human environment.

These women were not feminists as we understand this word today. But their world was not confined to house and home. They daily lived in a setting that encompassed both their home with their husbands and children and a surrounding natural and social environment that threatened at any moment to overwhelm and literally bury them in rock slides, avalanches, moral degradation, and lawlessness. A supporting community was a life-saving necessity. They saw to it that it came into being. So they began with home schooling their and their neighbors' children in what was then known as a mining camp. With the support of their husbands they then began the struggle to promote, fund, and establish the public school house. They made sure that it was the first public building in their new town, before there were churches, a city hall or a county court house. In fact, the school house served all of these functions, except that it was not open for town meetings—the use of tobacco and foul language, common at such occasions, would not permit that. But the school house was always open for social entertainments—literary evenings, spelling bees, concerts, and dances. They built and solidified community, and that was what mattered. It was in subsequent years that Silverton women also assumed public positions that until then had always been occupied by men. In 1886 Silvertonians elected their first woman school board president and ten years later a woman was chosen as superintendent of public schools of what is now San Juan County.

The history of the school presents us with an intriguing story of the impact on the children's schooling of an unforgiving nature with its house-high snowdrifts and its avalanches in winter and its forest fires in summer. It relates the battles over school board elections and the often vituperatively fought newspaper debates between editors who favored progressive reformers and taxpayers who resented any dollar spent on the school. It shows the board's never-ending problems over the hiring and firing of principals and teachers. And as we cannot understand these events and controversies over educational matters without also being aware of the issues that affected the town and its people, we have to know of the town's history as well. We need to read about the ups and downs of the mining industry that constituted the town's economic life-blood. We need to hear about the townspeople's struggles with the social issues raised by nonwhite immigration, threats of attacks from neighboring Ute Indians, opium dens, prostitution, shootouts by gun-slinging desperadoes, and frontier-style lynchings. We have to be reminded of the West's and the nation's political turmoil over the free silver issue, labor unrest, the War with Spain and the debate over imperialism. Only thus can we understand how with the graduation of its first high school class in 1902 both school and town jubilantly greeted the new century.

For the next 100 years town and school proceeded on their chosen path of mining the precious ores and preparing its children for their adult lives. But when by the end of the twentieth century mines began to close one by one and Silverton's population precipitously declined, the school was threatened with closure. In a postscript I tell how when after the closing in 1991 of the last producing hard rock mine Silvertonians reacted to the threatening disaster by adopting in 2002 the Expeditionary Learning—Outward Bound (ELOB) curriculum. An unconventional academic and experience centered form of progressive education, ELOB had developed from its beginnings in the 1920s in Kurt Hahn's *Schloss Schule Salem* in the Swabian Hills on Lake Constance. It had found its way via Scotland and Wales in the 1940s to the mountain fastness of the San Juans where it today governs the educational philosophy of the Silverton public school.

I should add a word about this book's place in the writing of the history of the American West, the history of education, and women's history. As a subject of historical interest the West first attracted the attention of Americans when in 1893 Frederick Jackson Turner proposed the frontier thesis of American history. Turner claimed that the westward expansion of settled land and the battle fought there between civilization and savagery shaped American character, confirmed the belief in manifest destiny, rugged individualism, and democracy as a form of government. For 100 years this was the preferred theme of Western historiography, with its chapters on Indian wars, mineral exploitation, the fur trade, violence, and the accounts of colorful and adventurous characters in forbidding terrains.

Today, however, historians of the American West, while not entirely forsaking these topics, have come to favor a different approach. They write about the contributions of Native Americans, Latinos, Afro-Americans, immigrants from Asia, and women. They discuss the impact of settlement on the environment and discuss social and religious developments.

But few have paid special attention to educational developments in the Mountain West. It is in this context that we can gain new insights into the history of American education and the history of women in this country. There is first the simple fact that, when compared to New England or the Midwest, the Mountain West has few histories of its educational past. We know a good deal of its colorful and adventurous characters, their exploits in an often forbidding terrain, but there are few accounts that tell us how the families of explorers, miners, and settlers dealt with the education of their children and how public education gained a foothold in the wilderness. This book will help fill this gap.

Second, while we have known that women have played a contributing role as "helpmates" to the male school reformers of the nineteenth-century public education movement in many parts of the country and that they have been true and effective pioneers in women's higher education, in Silverton we meet them as "makers and shakers" of public education. Here they fought their battles on their own turf relying on their own resources.

What made their efforts so remarkable was that they managed to persevere in their pioneering while balancing it with the conflicting impulses of their own upbringing's belief in the woman's sphere of home and family, impulses that urged them to be gentle, sweet, and nondemanding. It was their daily experience of living in a rough, crude, and unforgiving natural and human environment that forced them to assert themselves and take on positions as promoters and teachers of public school education.

These insights also help to explain the so obvious contradiction between the curriculum the first high school graduates had been given to study and the employment choices they made. For our women school sponsors high school graduation meant a portal into a world away from home, to college attendance and a professional career. That is why they and the school directors introduced the Latin-Scientific course that permitted entry into the state university without examination.

They also knew that the alternative, a work-preparatory course, would seem superfluous to most Silvertonians and thus was never considered. Silverton's first four high school women graduates, however, combined their liberal arts education with an introductory so-called normal course that prepared them for teaching in a public school. They did not share the school women's commitment to equal professional status with men, but for a few years applied their liberal arts education to teaching in Colorado's public schools.

The third new insight that we gain from this study is that the German contribution to American public schooling is not, as we usually hear and read, a matter of the past but continues to this day. While the first wave of German educational influence, the Pestalozzian enthusiasm for a benevolent teacher paternalism, had engulfed the public school movement of the 1830s and 1840s before white settlement had reached the Rocky Mountains, the second wave, the vogue of Herbartianism with its five steps of instruction and its theoretical concepts of apperception and culture epochs, swept America's teacher training institutions in the 1890s and thus had reached the West. But what is less well known is that other modern progressive school practices have crossed the Atlantic Ocean during the twentieth century. For Silverton this concerns the introduction in 2002 of the Expeditionary Learning—Outward Bound curriculum, an educational practice that traces its origins back to Kurt Hahn's co-educational boarding school *Schloss Salem* in Swabia on Lake Constance. In the book's concluding chapter I describe the *Schloss Salem* educational experience, how it was transferred to Scotland and Wales and there turned into Outward Bound, and how in the United States it subsequently emerged as Expeditionary Learning.

As a fourth new insight I suggest that a proper understanding of this latest importation of German pedagogy will also make us aware that progressive education in the United States has not only been, as is commonly asserted, a by-product of American political progressivism, but a part of a worldwide countermovement within state-established public education systems. It has been a lively and effective protest to administrative educational bureaucracies and to pedagogical rigidities in classrooms everywhere.[1] In Silverton it was not introduced as a privately sponsored venture but as the guiding philosophy for the town's public school.

Thus these four insights give us a new and fresh understanding of American educational history. They underscore the strength of the American commitment to public education as a mainspring of the nation's vitality.

As I look back on my labor of love to reconstruct the story of this mining community, its school, and its valiant citizens, I have only one regret. It is that neither Freda Peterson, the archivist of the San Juan County Historical Society, nor I succeeded in finding memoirs, correspondence, diaries, and other personal recollections of school board members, teachers, students, parents, and others. Thus I could report little of what lay behind the incessant struggle of school board members over the hiring and firing of teachers, of the hopes and disappointments of teachers, and of the experience of students and their parents as they witnessed the daily life in the school's classrooms. As a historian, well aware of the value of private documents for historical research, I can only express my regret over their absence.

Thus the story I tell focuses primarily on the political fortunes of public schooling as I could unearth them from the newspapers, school board minutes, and other public documents of the period. It affirms as its main assumption that the history of an institution, be it school, church, or public library, cannot be fully understood unless it is placed in the general economic, political, and social development of its locality. Thus the chapters of this book are concerned equally with Silverton's business, politics, and social life as the background against which public education should be seen.

*   *   *

It is now my pleasant obligation to express my gratitude to the many individuals and institutions who so generously offered help and advice during my research. My thanks go to the Ballantine Family Fund for its much appreciated financial support; to the staff of the Archives of the South West Center of Fort Lewis College, above all to Andrew Gulliford, the

Center's then director, and Todd Ellison, its then archivist; to Freda Peterson, the archivist of the San Juan County Historical Society, and to Beverly Rich, the Society's president; to Chris George, school board president and Donna George and her successor, Tracy Boeyink, business managers of the Silverton school, who gave me full freedom to explore the school's archives; to Kim White, superintendent of the Silverton Public School District No. 1, and her secretary Evelyn Archuleta, all of whom made working in the school's archives a pleasure. I much appreciated the help I received from Jackie Leithauser in searching school census records in Silverton's Public Library, from Jonathan Thompson, then editor of the *Silverton Standard & the Miner*, and from Phyllis Kroupa and her staff at the Interlibrary Loan Department of the Reed Library at Fort Lewis College in Durango.

One of the special pleasures in pursuing my research has been the opportunity of being able to rely on the work of earlier historians. For anyone who is interested in Silverton's history in the 1870s and early 1880s there can be no more stimulating experience than perusing the vivid and searching narrative of Allen Nossaman's superb trilogy, *Many More Mountains*. Its first volume, *Silverton's Roots;* the second, *Ruts into Silverton;* and the third, *Rails into Silverton* are treasure troves for any historian who would not only want to get the "facts" and the "data," but also the "feel" for the drama of the events Allen describes. As a newspaperman of long-standing, Allen has based his story on the town's newspapers, and I have tried to follow him in this endeavor.[2] Of inestimable value in identifying the major actors in Silverton's history has been Freda Peterson's admirable publication, *The Story of Hillside Cemetery, Silverton, San Juan County, Colorado*. It serves as a splendid guide to the lives, trials, and triumphs of Silvertonians and is a true monument to Freda's outstanding diligence and accomplishment as archivist and historian.[3]

There remains for me now only to convey my most deeply felt and most happily expressed confession of love and gratitude to my wife Susan who throughout all these years has unfailingly given me her advice, encouragement, and support in my endeavors as historian and author.

CHAPTER 1

BEGINNINGS

Too much, sir, cannot be said of this country.... Those who come simply to look, remain to work. Some get homesick, others get broke, but all unite in testifying "It is the richest mining district ever discovered."

—From the first issue of the *La Plata Miner*,
published at Animas Forks on July 5, 1875,
reprinted in the *Miner* on December 30, 1882

It was Sunday morning, September 19, 1875, in Silverton, La Plata County, Territory of Colorado. The inhabitants of this silver mining camp in the middle of the big meadow called Baker's Park, surrounded on all sides by the snow-capped peaks of the San Juans, gathered at their newly constructed school house. They had come to celebrate the beginning of public education. During the preceding months, La Plata County's superintendent of public schools, Mr. Jacob M. Hanks, had persuaded the parents of twenty-eight children to enroll their boys and girls. That was a considerable achievement in those days when to enforce compulsory schooling was still considered an impossibility and many families left town over winter to escape being cut off from the outside world by snowdrifts and avalanches.[1] The twenty-eight children Hanks had mentioned in his report to the territorial Superintendent of Public Instruction represented no less than 72 percent of Silverton's youngsters between the ages six and twenty-one, the group at that time considered to be of school age.[2]

But Superintendent Hanks did not play an official role in the morning's proceedings. Hanks, who in the previous year had carried out a preliminary survey of what was to become Silverton, had been nominated as superintendent and had run unopposed in the county's first election on September 6, 1874. There not being a school house in La Plata county, he lived at that time in Del Norte, ninety miles to the east of Baker's Park across Cunningham Pass on the Continental Divide. There he stayed for the winter, teaching thirty-one children at the West Del Norte school. But when on February 1 of 1875 Silverton was named as the seat of a U.S. Post Office and Hanks was named the postmaster, he immediately left for Silverton to assume his new duties. In Silverton, however, there was no building available to house the new post office, and Hanks had to improvise his postal service. He succeeded so well that Allen Nossaman, Silverton's premier historian, could report that "Hanks gained both the admiration of his townspeople and legendary status by running the delivery aspect of the post office out of his hip pocket." A letter would stay there until Hanks met its recipient on the street. It was not until July that he was able to open a post office where, besides postal items, he also sold sundries and newspapers.[3]

Hanks had other irons in the fire as well. Experienced in the printer's trade, he helped bring out on July 11, 1875, Silverton's first newspaper, the *Silverton Weekly Miner,* and in the next year he functioned for a while as its editor. During 1875/1876 he also served as a justice of the peace, in 1876 acted at least once as coroner, and in the winter of 1876/1877 took over the duties of county treasurer and county clerk. Concerned about safe and reliable postal service with the outside world that, he knew, was not guaranteed by Barlow & Sanderson, the stage coach company commissioned by the U.S. Post Office Department, he was instrumental in organizing the San Juan and Silverton Turnpike Company whose incorporation papers he signed as secretary in January of 1876.[4] Jacob M. Hanks was a busy and well-liked man in Silverton, and his many other duties may well have kept him from attending his school's September dedication.

The Superintendent of Public Schools not being available and the town not counting among their residents an ordained clergyman—nor, for that matter, having as yet a church building of some kind—the task of properly inaugurating public education fell to the town's business community. One of their leading members, Mr. W.W. Perkins, partner in Silverton's Rough and Ready Smelter, opened the proceedings with a sermon. When he had finished, another partner in the Rough and Ready Smelter enterprise, Mr. B.F. Holmes, stepped before the assembled crowd and invited all to stay and share with him the lesson of the local Sabbath school. In this mountain fastness, 9,318 feet above sea level, and, apart from an occasional visit by groups of passing Ute Indians, isolated from the rest of the world, public education Western style made its entry under the sponsorship of Christian religion and Western business enterprise.[5]

The opening of the public school in Silverton occurred in the fall of the town's second year of permanent settlement. It followed a period of rapid mineral prospecting, consolidation of mining claims, and establishment of industrial enterprises in the mountains surrounding Baker's Park. The town's lifeline in those early years before toll roads and railroads were built was the ninety mile pack and wagon trail that led along the Rio Grande from La Loma, later transformed into Del Norte, past Wagon Wheel Gap through Antelope Park, Timber Hill, and Grassy Hill across the Continental Divide on Cunningham or Stony Pass to Howardsville.[6] Men, driven by the lure of gold and silver and the thrill of exploring mountains and planting home sites, willingly faced getting lost in the wilderness, accidents, and starvation, and in winter death by avalanche and freezing. Once, however, the first settlements were established and wives and children joined the early explorers, it was time to think of schools and churches. Culture and civilization were to arise side by side with the turmoil of economic exploitation and development.

Legal recording of what would eventually become part of San Juan County had begun with setting up the Las Animas

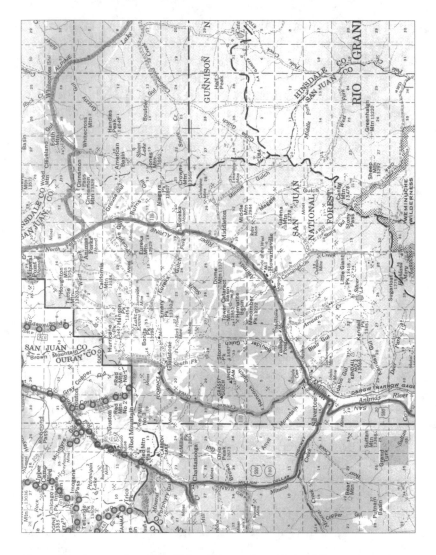

Figure 1.1   Excerpt from Travel Map, Uncompahgre National Forest, Colorado

Mining District on June 15, 1871. It took place in prospector Miles T. Johnson's tent near Arrastra Gulch three miles north of Baker's Park. In the previous year, Johnson, together with Dempsey Reese and Adnah French, had there discovered the Little Giant lode of gold and silver ore. Johnson recorded it in July of 1871. In those proceedings he acted as deputy clerk of Conejos County, the county that then covered most of the San Juan Mountains. When winter approached prospectors left the area, and many withdrew to Del Norte, a settlement that at that time consisted of between twelve and fifteen cabins. They returned the next spring, built the first log cabins in Arrastra Gulch near the Little Giant, and brought a quartz mill across the Continental Divide over Stony Pass. The mill, however, did not begin its operations until June of 1873. Six miles further north at Eureka other prospectors established a new mining district, and at the confluence of Cunningham Creek and the Animas River George Howard erected the first permanent cabin. Soon to be known as Howardsville, that place served for a short while as county seat. In 1873 Henry F. Tower, accompanied by his young wife and eight bull whackers, had three ox teams pull a sawmill weighing 6,000 pounds from Colorado Springs to Howardsville. The party crossed the Rio Grande fifty-three times and then had to build a four-mile road from Howardsville to Baker's Park where the mill was put up at Mineral Creek. Together with Dempsey Reese's cabin, the mill became one of the first buildings in Baker's Park. When in the spring of 1874 Jacob Hanks and others carried out a survey of what was to become the town site of Silverton, they showed approximately 100 residents and twenty-five cabins.[7]

During all this time, governmental and legal events had impinged heavily on local developments. Prospectors and miners who had felt unjustly hampered in their activities by federal threats to enforce Indian treaty rights and prevent them from carrying out their explorations felt relieved when in May 1873 the government withdrew its restrictions. An agreement had been reached with the Utes and their Chief Ouray that was ultimately ratified by the U.S. Congress on April 29, 1874. In the same year the Colorado Territorial legislature broke up Conejos County, and the Las Animas Mining District as well as Baker's Park found themselves in the newly created La Plata County with Howardsville at its seat. Residents of Baker's Park, however, did not relish that arrangement and took matters into their own hand. At the September 1874 election, the first to be held in Baker's Park and the one in which Jacob Hanks had been chosen as county superintendent of schools, the voters expressed their preference for Silverton as the county seat by a vote of 183 to 102, with Eureka with its fifty-one votes running a distant third. The centennial year of 1876 then brought the final governmental changes for Colorado and the San Juans. The Territorial Legislature split La Plata County in two. The northern half with Silverton at its seat became the County of San Juan on May 1. Three months later, on August 1, the Territory of Colorado attained statehood.[8]

By the fall of 1874 Silverton had asserted its dominance as the population center of what was then still La Plata County. As winter drew near many of its inhabitants decided to stay for the duration. So it was that Allen Nossaman could write that "Silverton assumed a mantle of permanence during 1875." In that year "between 80 and 100 new structures were erected...and virtually all of them were of frame construction, contrasted with the log building which characterized structures the previous year." The shift from log to frame construction had come as a result of Henry F. Tower's sawmill that had begun operations in 1874. By September of 1875, the very month in which the school opened its doors, the *San Juan Prospector* estimated the population to have reached about 1,000. However much exaggerated that figure may have been, Silvertonians could look forward to the future with confidence.[9]

The opening of the first public school house in Baker's Park in 1875 had not been the beginning of schooling in Silverton. For that we must look to the settlement's first families who began what we today would call home schooling in a two-story residence on Reese Street built in 1874 by George Greene, the financier of Silverton's Greene smelter and store. Three families lived in the house: George Greene's son Edward Merritt Greene with his wife Emma Eberhart; William Edward Earl and his wife Julia Emma Marsh with their ten-year-old daughter Artie and four-year-old son Frank E., and Benjamin W. Harwood, trail blazing freighter, with his wife Sarah Elizabeth Spillman, their two-year-old daughter Mary Belle, and her one year old sibling Barbara. By the summer of 1874, the families living in the house began teaching their own children and, most likely, others as well.[10]

Instruction of children was closely tied to religious observance, and the first teaching was done in the Reese Street house by Julia Earl and Sarah Harwood who was recognized as a "pioneer in the fundamentals of the moral and social fabric of Silverton."[11] Schooling for the children as well as Sunday morning church services also took place at the Eatons' residence on Reese Street. There the gentlemen of the Rough and Ready Smelter, W.W. Perkins and B.F. Holmes, as well as John and Susan Eaton presided over the activities. The Eatons, affiliated with the Gospel of the Free Will Baptist Church, were also connected with the Rough and Ready Smelter. Besides the three wives already mentioned the *La Plata Miner* praised others in Silverton for "the pluck to remain in this isolated region as the companions of those they love" and for having been "made of the same kind of stuff as were our fore-mothers of a century ago..."[12] They were Amanda (Mrs. John F.) Cotton, who probably was the first woman to settle in the area, Miranda (Mrs. John) Lambert, who when she died in 1933 was the oldest woman resident, Carrie E. (Mrs. William H.) Nichols, whose son was the first to be born in the county, and Emma (Mrs. Edward) Greene, Anetta (Mrs. Hans) Aspass, Mrs. George Walz, Mrs. B.F. Holmes, Mrs. Harvey L. Rodgers, and the Earls' daughter Artie.[13]

These women were greatly outnumbered by the 330 men who cast their votes in the election on September 6, 1874, and, barred by law from the ballot box, could not have contributed to the political destiny of Silverton in this way. But their influence was far greater than numbers would indicate. The men in the community, to their credit, readily acknowledged that, as one of them later reminisced when writing of the winter of 1874/1875:

> ...for genial kindly feeling and sociability, that winter has never been equaled, and never will be, in the metropolis of the San Juan. Because we were few and isolated, each and all tried to make the best of it, and as pleasant as possible for the rest. Our number of ladies was small, but "select," i.e.[,] for their ladylike bearing, their good nature and sociability.

The "select ladies" saw to it that schooling, religious services, and social events such as dances and debates kept the community together and mutually supportive. Although they all enjoyed a good time, alcoholism among the men was kept within bounds. One of Silverton's menfolk, prospector Charles H. Slocum, paid the Silverton women what was perhaps the highest honor: he staked out a claim in their names and called it "the Rose of San Juan."[14]

Nevertheless, winter was a trying time for all of the Silverton residents. Two feet of snow fell early. The cabins were soon completely covered, and tunnels were dug between them. Work in the smelter became intermittent. John F. "Jack" Greenell, their sheriff, acknowledged the hardships, and tried to be as helpful as he could. This wasn't always easy for him. As he wrote to a friend:

> We have some very desperate men here, men that think nothing of taking a man's life. Shooting and cutting is the order of the day. As soon as the lie is given, revolvers are drawn and one man passes in his checks.

But when the snow cut off Silverton from the rest of the world, Greenell and his neighbor Hans Aspass put on their skis, then called snowshoes, and three or four times skied the ninety miles to Del Norte and back to fetch mail and provisions. Back home in Silverton, Greenell showed his appreciation to the ladies by giving them skiing lessons.[15]

But Silverton's women did more than teach home and religious school and foster community spirit through social events. During the summer of 1874 Julia Earl and a Miss Norton, of whom little more than her name is known today, had begun a fund raising campaign for the public school house that was to be inaugurated on that September Sunday in 1875. They and their other "select ladies" were the first promoters of public education. But they did not see themselves as pioneer feminists. They remained attached to what was called the women's sphere in house and home. The reasons are not difficult to see. They sprang from the backbreaking physical labor that mining demanded and the ever threatening physical and moral environment—mine disasters, avalanches, snow blockades, forest fires, threats of Indian attacks, and prostitution, opium dens, shootouts and lynchings—that made them care equally for their homes and their community. They knew they had to care for their menfolk and children at home and prevent them from descending into savagery and crime. So they became "culture bearers" and strove equally for harmony and literacy at home and, in the public school house that they promoted, for education, literature, art, and sociability in town.

Though visiting Baptist and Methodist ministers from Del Norte had held religious services in Howardsville in July of 1874, there still lived no resident clergyman in Silverton during the winter of 1875/1876, and laymen continued to preside over religious services in the town's private homes. Thus it seemed natural that the women fund raisers would conceive the new structure to serve the community as school, church, social center, town hall, and court house. This did in fact happen during the next few years, except that municipal government was never permitted to occupy the building. The use of tobacco and foul language common at town board meetings were thought to violate the sanctified atmosphere of the school.[16]

When the public school house opened for classes in September of 1875 the existing private schooling did not immediately disappear. The county's superintendent, Jacob M. Hanks, reported that, despite this competition, he had succeeded in organizing one school district and enrolling twenty-eight children.[17] According to both Allen Nossaman and the *The San Juan Prospector* of September 25 and October 23, 1875, Mrs. Matilda Huff served as the first secular teacher throughout the winter of 1875/1876, having fifteen children in her class.[18] Apparently, not all of Hanks' twenty-eight students had shown up. This, as we shall see, was to become a recurring problem. Attendance numbers at school never were to equal those of the census or those of the school's enrollment statistics.

CHAPTER 2

A HOUSE OF MANY USES

In all America there was, perhaps, no better symbol of the shared community life people remembered than the one-room schoolhouse standing in the center of an independent school district...

—Wayne Fuller, *The Old Country School*, p. 60

July 1875 had seen the beginning of construction of Silverton's school house at the corner of Snowden and Eleventh Street. When finished in September the building, a one-story big hall frame structure, measured twenty-five by forty feet. Its front door with a window on either side of it faced Anvil Mountain. There were two windows each on its north and south side, and a peaked roof topped off the structure.[1] Benjamin Wilson Marsh, a freighter and charcoal burner, wrote at the end of July that the approximately seventy-five houses of the town included homes with enough children to fill the school once it was ready. And some of Silverton's and the county's residents did their part to ensure that there would be pupils in the future. Byron Parker Taft and his wife Sarah Elizabeth Morgan Taft, sometime keepers of a boarding house, were the parents of Silverton's first baby girl, Anna Silverton Taft. Born on July 29, her middle name was to make sure that the place of her birth would be remembered as long as she lived. Just seven days later, on August 5, Silverton's second baby, Nevada Ann, was born to Mason Farrow and his wife Martha Jane Miller Frasier Farrow.[2]

Also in July 1875 Silverton's *La Plata Miner* issued its first two editions and could report that Otto Mears, known in the West as "the Pathfinder of the San Juan," had arrived to build the fifteen-mile road from Silverton to Mineral Point via Howardsville, Eureka, and Animas Forks. If parents in these outlying settlements so chose, their children then could more easily reach Matilda Huff's classroom in the Silverton school.[3]

But of even more immediate significance for the school's opening in October was the wedding that had taken place in July across the Continental Divide in Del Norte. Matilda Brown then had become the wife of David Huff, a Silverton miner and carpenter. The young and inexperienced teacher, however, did not fare well in both her pedagogical and personal affairs. By the time three to four feet of snow covered Silverton's streets, Matilda Huff found it ever more difficult to keep her fifteen charges under control.[4] Her teaching apparently was not a resounding success. She could not hold the boys and girls of varying ages busy on their seats when outside the school house door the 10,900 feet high eastern ridge of Anvil Mountain with its steep and snowy slope beckoned for sliding and snowball fights. Then, too, her marriage had barely lasted three months when David Huff departed Silverton and left her to cope with her troubles by herself. Timothy Parker Plantz, who filled in as her medical adviser, commented in spring that she was "still suffering by overdoing herself." So when school closed in April Matilda Huff turned in her resignation.[5] Her successor, William K. Newcomb, arrived from Del Norte where, just like Jacob Hanks before him, he had taught at the West Del Norte school. In Silverton he supervised Matilda Huff's fifteen pupils for the three-months summer term of 1876. But as the school directors were unwilling to finance school for another winter he, like his predecessor, took his leave.[6] For the Silvertonians, however, these were minor problems. They were happy to have a school house of their own, even if that building was not always in use and had to serve more than one function.[7]

During the winter of 1875/1876 around 200 souls of the twenty-five families and assorted bachelors remained in town. As the weather would permit, they stayed in intermittent contact with the miners and handymen who lived in the surrounding mountains and continued to work on their claims. Avalanches exacted their first victims from among them. As the new year began Silverton's fifty buildings included two general stores and one drug store, several hotels and restaurants as well as at least five saloons and two dance halls; blacksmith, butcher, shoemaker, and carpenter shops; a bakery, a barbershop, and a brickyard; a newspaper and a post office, four lawyers' offices, a sawmill, a livery stable, the town's major industrial plant, the Greene Smelter and, of course, the school house. But the town still lacked a bank, a church, a city hall, and a county court house, a permanent full-time physician and a resident clergyman.[8]

To the chagrin of many of Silverton's respectable citizens and, no doubt, to the delight of the miners living near their claims and coming to town for occasional visits, prostitution had made its appearance by October 1876. "Fancy girls" were reported to live on lower Greene Street. William Bowen with his English-born wife Jane subsequently opened their Westminster Hall on Blair Street. Jane, who had grown up in London's Petticoat Lane neighborhood, came to be known in Silverton as "Aunt Jane" or "Sage Hen." As the town's first Madam she presided over such "soiled doves" as Annie Norman, Belle

Figure 2.1　Silverton, Notorious Blair Street

West, and Maggie Davis. Thus it came as no surprise when at a town board meeting in the following May two petitions were introduced, one urging the board to restrict "dance houses and houses of that class," the other asking to either suppress them altogether or "to take no notice of them." A vote was taken on the first motion and as it resulted in a tie, no action followed. By the spring 1877 prostitution had gained a firm foothold in Silverton.[9]

The nation's centennial year not only witnessed on August 1 the admission of Colorado as the thirty-eighth state of the Union but also brought several important political events to Silverton and San Juan County. On October 3 Silverton's voters went to the polls and elected William Munroe both county surveyor and county superintendent. They would do so again a year later and reelect Munroe for two-year terms ending in 1879. His predecessor as school superintendent, the indomitable Jacob Hanks, had not run for reelection in 1876 and was to leave Silverton in September 1878. On November 15, 1876, the settlement in Baker's Park officially became the Town of Silverton when the county commissioners endorsed its incorporation. San Juan County, however, was cut down to its present size by the State Legislature in January 1877. The county's territory lying north of the Animas River and Poughkeepsie drainages was separated and reconstituted as the County of Ouray.[10]

The centennial year also brought new luster to Silverton's school house and enhanced its importance as the town's major cultural center. As before, it was home for church services led on Sunday mornings by Silverton's leading lay citizens. On Sunday evenings it served the Silverton Literary Society as its meeting place for weekly debates and lectures. Afterward participants would gather at the Cotton boarding house and residence where their hosts would play for them, John Cotton on his violin and Amanda on her melodeon. There they could celebrate the birth of Silverton's only child to be born that year in the San Juans. She was Kittie Clark, born on the last day of July, the daughter of Giles Perkins Clark and Dorothy "Mizzie" Cameric Clark. While the "cultured" citizens of Silverton danced, played cards, had their fortunes told, and thus enjoyed themselves at the Cottons', those less attracted to literature and family entertainment found their way to the two bars on Greene Street. The Canadian and the Gem made sure their patrons could find there the pleasures they sought.[11]

The school house served other functions as well. On the first day of July 1876, a Saturday, it became the polling place for the county election at which the state constitution was adopted. The next day a visiting Boston clergyman, the Rev. F.B. Perkins, officiated at an evening church service. Perkins, so Allen Nossaman surmises, "was the first professional clergyman to conduct a service in Silverton..." Folks then got ready for the July Fourth celebration on the following Tuesday. After a rally on Greene Street the marchers crowded into the school house where Superintendent Hanks read the Declaration of Independence and attorney Columbus W. Burris delivered an address fitting the occasion. After horse, burro, and foot races and a baseball game in the afternoon the school house beckoned for a grand ball in the evening. Amanda and John Cotton with their melodeon and violin provided the music.[12]

The uses of Silverton's school house received official recognition at the town board's meeting in April 1877. Some members urged that the Board should have access to the school house that they thought should more properly be called Town Hall. The defenders of propriety and upholders of the building's role as church and school, however, protested and prevailed. Nothing came of the matter. In fact, the use of the building for church services held by visiting clergymen became routine. In late June Baptist Rev. A.B. Whitney preached in the school and on July 22 Colorado's Episcopal bishop John Franklin Spalding presided over three services. At about the same time Father Thomas A. Hayes conducted a Catholic mass, although it is not sure whether he did so in the school house or in some other hall. On August 24 Silvertonians chose the school house for a decidedly nonparochial event. They threw a community benefit for the Moyle Brothers Brass Band, a family of Cornish miners who worked their claims on Sultan Mountain and had entertained their fellow townsmen on many an occasion. Just as at the July Fourth ball of the preceding year, music and hilarity, dancing and merry-making seem not to have offended the upholders of sanctity and propriety. What they found objectionable were the sound of foul language and the squish of tobacco juice.[13]

In May 1877 Silverton's school district, originally organized by Jacob Hanks, carried out its first election. It chose John L. Pennington, George Walz, and William Edward Earl as its directors. All three belonged to the town's business elite and had been in various ways associated with the Greene enterprises. Walz, who like Earl had lived in Silverton through the winter of 1874/1875, was a mine developer and had moved with his family to a cabin at the southern edge of town on Sultan Mountain. Earl's wife, Julia, had been one of the early home teachers in the Greene residence and had promoted the funding of the public school. Pennington, who had arrived early in 1875, served as manager for the Greene store and smelter. During the winter of 1875/1876 both he and Earl had taken leading roles in the town's social affairs. Pennington appeared in the Literary Society and lectured on "Recreation," while Earl chose "Slander" as his topic. No doubt, their much-applauded presentations convinced voters that they were ready for their jobs as school board members. Their first action then was to appoint a successor for William K. Newcomb. Their choice fell on nineteen-year-old Huldah Ann Puckett, just arrived from Kansas with her parents and six siblings. Four of them, two sisters and two brothers, were of school age and became pupils of their older sister. The town's third public school teacher began her duties on June 11. At the end of the term in the fall she returned to Kansas to get married to Alfred T. Jardon. Though the newlyweds came back to Silverton in the following spring, Huldah chose not to resume her teaching.[14]

Again the school board had to search for and appoint a new teacher. This time the choice fell on Helen Josephine Standish Bowman, the wife of the Greene Company metallurgist, Thomas Evans Bowman. Helen Bowman agreed to teach for the summer term of 1878. She entered upon a mutually satisfactory working relationship with the Rev. Harlan Page Roberts. While she struggled with her around thirty students during the week, Roberts, a recent graduate of the Yale Theological Seminary, preached to Silverton's Protestants on Sundays. By the time school closed at the end of the summer, Roberts had achieved what Silverton's businessmen and their wives had so desperately hoped for. He had united the various Protestant groups of the town. On November 24 he could proudly report the founding of the Congregational Church of Silverton to which he would continue to minister in the school house. As in previous years, there also appeared visiting Episcopal and Presbyterian ministers to preach their sermons, and Father Brinker celebrated an occasional Catholic mass.[15] Helen Bowman, on the other hand, considered her obligations completed and resigned from her position. Members of the school board who had just decided to initiate a regular fall term for 1878/1879 felt themselves lucky to be able to persuade the twenty-two-year-old Emma Hollingsworth, youngest daughter of attorney Leander F. Hollingsworth, to take on the assignment.[16]

For the superintendent and the members of the Silverton school board the task of keeping a public school in operation during those first four years must have been perplexing, annoying, and ultimately rewarding. Perplexing, because their own and their fellow Silvertonians' struggle for physical and economic survival in an unforgiving environment must often have seemed far more important and demanding than any call for the care of a building whose use remained forever disputed; annoying, because of the ever-recurring search for candidates who could be persuaded to act as a teacher; and yet rewarding because, over and again, the school shone as a beacon of light in the wilderness.

There can be little doubt that the town women's efforts at promoting and fighting for the school had succeeded and had gained it a foothold because it had received the unflagging support of its first superintendent, Silverton's early postmaster Jacob M. Hanks. His fellow citizens trusted him. They knew that what he undertook was for their own and their town's greater good. At the same time, Hanks' many and varied other activities showed all too clearly that a school superintendent's job, however important and beneficial for the community, did not take up a man's whole time and attention. There were other things to be done for the community, such as safeguarding the mail and upholding law and order. Hanks knew how to weigh one against the other. He had no patience with bureaucratic and administrative requirements that showed no understanding of a pioneer settlement's conditions. He wrote the territorial superintendent of public instruction that he had "no official and full report to make. I have made every reasonable effort to organize schools in this county, but owing to the nature of our country—the major part of the people leaving every fall—I have made but little progress." He hoped, he added, "to be able to make a good report next year." He never submitted such document.[17]

The state superintendent's reaction to Hanks' labors—and no doubt to those of other county superintendents as well—tell us that the ever-recurring frictions between the proponents of local and of state control of public education began at an early date.[18] In his report for 1878 State Superintendent Joseph C. Shattuck complained about "the careless, unbusinesslike way in which district accounts are kept." They caused the district reports to be "incomplete and inaccurate as to be quite unreliable." The statistics for San Juan County given in the state report for 1878 are a case in point. Obviously referring to Helen Bowman and Emma Hollingsworth as the two teachers for the year, the listing made it appear as though the school employed two teachers at the same time when, in fact, it had employed one each in each of two different sessions. The state report also falsely claimed that the school house was divided into four classrooms and that there were two school districts in San Juan County, the second being Animas Forks, supposedly established in 1881. That, however, was contradicted by San Juan County superintendent Dr. J.N. Pascoe in State Superintendent's Shattuck's Fifth Biennial Report. In 1887 San Juan County had only one organized school district.[19]

Of interest also is Mr. Shattuck's assertion that compulsory schooling in America was "a well proven failure," that it had been tried in eight or ten states but had there been found to be impossible to enforce. He concluded that "*education can not be made compulsory in the United States.*" There he echoed Jacob Hanks' complaints about parents taking their children out of school when they left for the winter; a practice that led the Silverton school directors for three years in a row to open their schools only in the summer.[20]

But Shattuck also had good news to relate. He praised the willingness of Coloradoans to pay female teachers nearly as much as their male colleagues, the average difference in monthly pay between men and women being a mere $ 2.95. In California, he showed, that difference came to $14.10 and in Connecticut to as much as $28.35. When he listed the average 1878 monthly pay for women teachers in Colorado as $46.95, he made the compensation for Silverton's teachers look quite respectable. Although we have no information about Matilda Huff's salary, we know that Huldah Puckett's monthly salary of $75.00 equaled that of Will Newcomb, though Helen Bowman and Emma Hollingsworth had to be content with $64.65. This was still above the state's average, but in the next year gender differentiation set in. Eber Smith, the teacher appointed for the summer of 1879, was paid $87.50. Apparently the school directors were becoming exasperated at the succession of one-term women teachers and hoped that with a higher salary they might entice Eber Smith to stay.[21] As it turned out, they were to be disappointed once more. Mr. Smith, too, became a one-term performer.

CHAPTER 3

TOWN AND SCHOOL IN A WILDERNESS

...a land that is paradoxically both extremely threatening and inhospitable on one hand and incomparably rewarding, inspiring and beautiful on the other...

—Allen Nossaman, *Many More Mountains*, vol. 3,
Rails into Silverton, Preface

As the 1870s drew to a close both town and school of Silverton had become firmly established in the economy of the San Juans. Its people liked to think of themselves as citizens of a mining metropolis whose men combined the rough and boisterous lifestyle of a frontier settlement with the amenities their wives provided through literary and religious services and their children's schooling. Silvertonians' lives were charged with the energy that drove their aggressive exploitation of the natural environment and were calmed by the efforts to bring culture and religion into their homes and public places. The seasons' rounds brought in winter the terror of burial by avalanche and death by freezing in the fearful whiteouts of snow and ice. In summer they brought the delights of glorious colors of wild flowers and aspen glens in mountain meadows and on alpine peaks. Silverton's people, miners and engineers, businessmen and lawyers, desperadoes and prostitutes, freighters and tradesmen, fathers, mothers, and children, experienced year-in and year-out the constancy and the immediacy of terror and beauty of life in the mountains.

Age made little difference. Whether adults or children, Silvertonians knew about the meaning of life and death. Daily they saw or heard of the vicissitudes of human existence in mine and smelter, shop and office, saloon and bordello, log cabin and two-story frame house. In that first decade of settlement, theirs was an experience of life in the raw. With the help of school and church, they were determined to make it liveable for young and old alike.

Access to Silverton remained a problem, in summer less so than in winter. As an outlet toward the northeast Otto Mears had finished by August of 1877 the construction of an extension of the Silverton-Animas Forks road for an additional twenty-five miles to Lake City.[1] Toward the south pack trails led on both sides of Sultan Mountain toward Cascade Creek and Animas City close to what would eventually become Durango.[2] By 1878 a toll road was being constructed through the Animas River Canyon. Not long thereafter it was learned that the Denver and Rio Grande railroad had bought it and planned to survey a route for a Durango-Silverton railroad up the Las Animas Valley.[3]

The preferred route into Silverton, traveled by most of Silverton's newcomers, remained for years the road that led from Del Norte along the Rio Grande past Timber and Grassy Hill across the Continental Divide over either Cunningham or Stony Pass past Howardsville into Baker's Park. Louisa Weinig, then a school-aged youngster, described the adventure in 1878:

> Our trip to Silverton was a wonderful experience. Tho the stage was crowded with passengers mother and I were the only women folks.... Our route was by Wagon Wheel Gap, the upper Rio Grande, Stony Pass and Howardsville. It took several days as we had to make over-night stops.... My brother and I and other passengers walked part of the way as the stage moved slowly owing to the steep and narrow roads. We enjoyed the beautiful scenery, so new and strange to us. Many mountain sides were still covered with snow, but flowers were blooming on the warmer slopes.[4]

Another experience was that of Peter Scott who for a while earned his living as a freighter. He tells us how with sixteen yoke of oxen he pulled his loads of provisions and machinery across the Continental Divide at an altitude above 12,000 feet. Once on top, the braking began, above the timberline with oxen tied to the rear of the wagons, holding them back as best they could, and once in the woods by snubbing. They tied ropes to the wagons and slung them around trees on both sides of the rocky path, the freighters holding on to the ropes with all their might.[5]

In winter, this route, perhaps more than any of the others, exacted a fearful toll in human pain and misery. Snow and ice threatened the traveler who on his longboards dared make his way across the Divide. It was there in November of 1876 that beloved Sheriff "Jack" Greenell met his death on Stony Pass as he carried the mail from Del Norte to Silverton. The following winter of 1877/1878 exacted a particularly high toll. Two miners were killed by an avalanche, and Samuel Greene, the nephew of the Greene brothers who ran Silverton's early smelter and store, froze to death on Grassy Hill.[6] The Rev. Harlan Roberts presided over Greene's candlelight funeral in the school house, and Greene's body, thawed out in a tub of ice water, was then taken at night to Del Norte to be sent to his wife in Cedar Rapids, Iowa.[7] The school house, too, served in May of 1879 as the meeting place for the formation of the Silverton and Grassy Hill Toll Road Company that was to ease travel between Niegoldstown and Grassy Hill over Stony Pass.[8] While the road by itself could not guarantee safety from avalanches, it eased and made more recognizable the route through the snow drifts that would guide the travelers across Stony Pass to their destination.

The fall of 1879, however, would remind the citizens of Silverton that it wasn't nature alone they had to fear. On October 11 their newspaper, the *La Plata Miner*, was full of reports and warnings that their Indian neighbors had engaged in battles with U. S. soldiers, among them a company of the colored Ninth Cavalry, and that the Utes were planning a general assault on all the white settlements in the area. The reports concerned the Ute ambush of Major Thornburgh's troops at Milk Creek and what came to be known as the Meeker massacre at the White River Agency.[9] Fears were increased when fires broke out in the woods surrounding Silverton and suffocating clouds of smoke drifted into town. As Freda Ambold told it later, "the smoke became so dense that mother had to fan the babies so they could breathe and the men decided to take their families in covered wagons and make a dash to get out before they were all suffocated." Rumors circulated that the fires had been set by Ute Indians. As it was, the rumors were never confirmed and a rain storm brought an end to the danger.[10]

Other changes, too, had left their mark on Silverton during 1878 and 1879. Some of them were viewed with grave suspicion by many residents, others were greeted with approbation and relief. The opening of two houses of prostitution on Blair Street raised eyebrows all around. Even though one of these bordellos burned down in August, rumors circulated among the solid citizens of Silverton that there still lived about a dozen soiled doves in their midst.[11] The town board subsequently passed an ordinance that made illegal the keeping of bawdy houses of ill fame, assignation, or gambling, and threatened with fines the hiring of prostitutes as bartenders who sang or danced in lewd manner. It was one thing to pass an ordinance, but quite another to have it enforced, especially so when many of the lawmen were the owners of the saloons and dance halls.[12]

The good news for Silverton was the arrival of Dr. Robert H. Brown and the appointment of Emma Hollingsworth as the town's teacher for the fall semester of 1878/1879. With Dr. Robert H. Brown Silvertonians now had for the first time a fully certified resident physician to take care of their bodily woes.[13] They also were delighted to hear that one of their own, Emma Hollingsworth, daughter of noted Silverton lawyer Leander F. Hollingsworth, was going to be their children's teacher for the four-month winter term. As presiding persona in the school house she would also be in charge of the traditional Christmas and New Year's entertainment. Not only the school children, but their parents as well, could look forward with happy anticipation to that event. They knew, they wouldn't be disappointed.[14] And Louisa Weinig had mentioned in her travel description that when she arrived in Silverton, Emma Hollingsworth was her teacher in the school house and also on

Anvil Mountain. There she supervised the children's sledding and taught Louisa how to ski.[15]

Throughout the calendar year of 1879 the Silverton school was much in the news. The school directors had selected Eber C. Smith as Silverton's teacher for the summer. As a graduate of the Kansas State Normal School at Emporia Smith was paid a professional salary of $87.50 per month, considerably above the $64.65 paid to Emma Hollingsworth, a daughter of the town and his predecessor. Smith also had been entitled to receive his teaching certificate without having to take the qualifying examination that William Munroe, San Juan County's superintendent of schools from 1877 to 1880, had given on the last day of May to candidates for a teaching position.[16] County superintendents were required by law to give such examinations quarterly to all applicants. Those who qualified were then given certificates of competency that allowed them to be paid from public funds for their teaching duties. The state superintendent reported that in San Juan County only one applicant had received a second grade certificate in 1879.[17]

By 1879 school census and attendance figures regularly appeared in the press and in the annual reports of the State Superintendent. They were, however, notoriously unreliable and fluctuated depending on the day they were taken and who had reported them. The *La Plata Miner* stated on July 12, 1879, that there were forty-four persons of school age living in town, school age then fixed between the ages of six and twenty-one. The county superintendent, always eager to boost his figures as much as he could, reported twenty-two boys and twenty-five girls for a total of forty-seven. Attendance, however, was a different matter. The Superintendent reported an average attendance of thirty-eight, but Eber Smith reported on July 12 that the enrollment actually had been twenty-eight, with an average attendance of twenty-four. He was proud that sixteen of the twenty-eight students had earned recognition for their attendance, deportment, and recitation record. Eight had been placed on the Star Roll and another eight on the Honor Roll.[18]

Despite his good record in the Silverton school and perhaps because of his Kansas State Normal School degree, Smith left the classroom and in September ran as candidate of the Greenback Party for the post of county superintendent of schools. He had learned, as had so many of his fellow normal school students, that the classroom teacher was the low man on the totem pole of the educational profession, that it was a position thought best suited for women, and that as a man he was to seek administrative positions of leadership. But he was defeated in the October election by the Republican candidate Henry O. Montague and left Silverton for Rico and a career as an assayer and journalist.[19] His successor in Silverton was James Edgar Dyson, a thirty-two year topographical engineer, who, while looking for a position in his field, accepted the offer to teach school during the fall and winter terms of 1879/1880. Being a college graduate, he had no difficulty either in obtaining a teaching certificate and was also paid a monthly salary of $87.50.[20]

One of the fall's great social events was the School Exhibition in the school house on Friday evening, October 3. Why the *Miner* wrote that "the school's children will have an exhibition," is not quite clear, because the children did no more than open the festivities with a song. Thereafter, "amateur talent" took over with what was described as a "quite lengthy programme." Teacher Eber Smith and two sisters of former teacher Huldah Puckett played roles in "Uncle Horace," a play about the rich relative from California whom his sister and nieces detest as a bore but love for his money. Apparently, the evening was a success. The *Miner* "respectfully" suggested

> that it would be a good idea for the young people of Silverton to organize a society for the purpose of giving a series of entertainments during the long winter now before us. We feel sure we have the talent here to make a success of such a society, and at all events it will do no harm to try.

The paper announced that "the entertainment with an entire change of programme" was to be repeated the following Monday night and the proceeds were to be used to buy an organ for the school.[21]

The *Miner* rather indiscriminately filled its pages with school tidbits whenever it struck the editor's fancy. Page fillers and anecdotes appeared as "femininities" and "news." Under the former we find: " 'Give me your hand,' said the school master sternly," raising his ruler to strike the maiden's fingers. " 'And my heart, too,' she replied meekly. Being pretty, her soft answer turned away his wrath." Under news we read: "An 18 year old school boy at Trinidad has committed the misdemeanor of hugging and kissing the school-marm when she attempted to flog him."[22] But there were also more serious items. In November the paper reported on "another one of those pleasant social dances at the School House." A week later it told the town that "the school house has a brand new blackboard, platform and other improvements for the beginning of the school year."[23] And when Christmas arrived it was Julia Earl who was in charge of the traditional Christmas Tree Celebration in the school house and it was Emma Hollingsworth, the former teacher, who read Longfellow's "The Famine."[24] As the 1880s began, Silverton's school house continued to be the center of the town's educational, religious, and social activities.

CHAPTER 4

A SETTLEMENT TAKES HOLD

> *Lord, how I loved the mining.*
> *There's nothing else I've ever done in*
> *my life I'd trade it for. To be called*
> *a miner, why it made me feel proud. For*
> *a miner could do it all. And he*
> *learned from the guys he worked with.*
> *They showed him the skills he needed.*
> *I'm proud I'm a miner.*
>
> —Billy Rhoades[1]

As the 1880s began Silverton and with it the mining region of the San Juans had been firmly established. As Allen Nossaman put it, town and area had been transformed "from a reticent wallflower to the belle of the ball.... Timbered portals, crudely mechanized shafts, sizable dumps and reasonable ore stockpiles" covered many of the surrounding mountains, gulches, and drainages. Add to this that by 1882 a railroad had reached Silverton from the south and that north of the town the discovery of huge ore deposits at Red Mountain had further boosted expectations of mining's success, Nossaman could write that "with its seniority among area communities and its development of religious, educational and civic facilities, it was as if Silverton had a pat hand, a straight flush, with the Denver & Rio Grande and Red Mountain as high cards."[2]

Table 4.1 1880 Federal Census of San Juan County

	Silverton	Mineral City	Pough-kepsie	Howards-Ville	Highland Mary	Precinct 4	Eureka	Grassy Hill	Precinct 3	Animas Forks	Gladstone	Total Outside Silverton
Inhabitants	514	71	76	98	72	26	14	20	76	44	76	573
Miners	167	39	52	57	66	20	1	13	67	36	53	404
Laborers	19	4	5	5	1	1	3	2	0	2	3	26
Merchants	17	1	2	5	1	0	0	0	0	0	0	9
Clerks	9	0	3	1	0	0	0	0	1	0	0	5
Carpenters	8	0	0	6	0	0	0	1	0	1	2	10
Engineers	7	0	1	2	0	2	0	0	0	2	2	9
Others	104	7	12	11	4	3	4	4	6	3	11	65
Employed Males	331	51	75	87	72	26	8	20	74	44	71	528
Wives	71	7	0	4	0	0	1	0	1	0	2	15
Single Females	13	0	1	0	0	0	0	0	0	0	0	1
Widows	4	0	0	0	0	0	0	0	0	0	0	0

Adult Females	88	7	1	4	0	0	1	0	1	0	2	16
Boys of School Age	19	2	0	0	0	0	4	0	0	0	0	6
Boys Age 5 or Younger	20	2	0	3	0	0	0	0	1	0	0	6
All Boys	39	4	0	3	0	0	4	0	1	0	0	12
Girls of School Age	26	5	0	3	0	0	1	0	0	0	3	12
Girls Age 5 or Younger	30	4	0	1	0	0	0	0	0	0	0	5
All Girls	56	9	0	4	0	0	1	0	0	0	3	17
Children of School Age	45	7	0	3	0	0	5	0	0	0	3	18
Children Age 5 or Younger	50	6	0	4	0	0	0	0	1	0	0	11
All Children	95	13	0	7	0	0	5	0	1	0	3	29

Source: Silverton Public Library.

Gold and silver ores and the miners who extracted it from the mountains surrounding the town, their peaks, drainages, and valleys were the building blocks on which rested Silverton's success. Miners toiled in mines and mills with exotic names such as Pride of the West, Old Hundred, Caledonian, Sunnyside, Mayflower, Mountain Queen, Gold King, Idarado, Green Mountain, and others. Seven out of ten of the 571 San Juan County miners enumerated in the 1880 Federal Census were housed at or near their workplaces in boarding houses in the settlements of Poughkepsie, Mineral Point, Animas Forks, Eureka, Howardsville, Highland Mary, Gladstone, Grassy Hill, and two unnamed election precincts. These 404 miners shared life in the mining camps with 124 other males, sixteen adult women, fifteen of whom were listed as wives and one as single, and twenty-nine children, seventeen of whom were girls and twelve were boys of school age or younger. Among the 124 men, we find common laborers, carpenters, engineers, assayers, blacksmiths, cooks, and freighters, representing occupations that were indispensable for the smooth workings of mines and mills. The others were listed as merchants, druggists, entrepreneurs, professionals, various tradesmen, clerks, and lawmen who supplied their families' needs.

In Silverton itself the census showed that 167 or every second of its 331 males was a miner. The other 164 males included nineteen laborers, seventeen merchants, nine clerks, eight carpenters, seven engineers, six printers, realtors, and lawyers and one law students, five druggists, four hotel owners, three bankers, one restaurant owner, one sheriff, one constable, one artist, one county commissioner, one physician, one postmaster, one minister, one county clerk, one probate judge, one teacher, one Chinese laundryman as well as blacksmiths, bakers, butchers, barbers, millers, painters, stone masons, and other tradesmen such as jewelers, cigar makers, liquor dealers, and sign writers. Side by side with these men lived eighty-eight adult women and ninety-five children. And there were the hangers-on, never-do-wells, and desperadoes, who flocked in and through the town. All together they made up a fascinating sample of courageous and footloose humanity.

Thus with 514 of the 1,087 inhabitants of San Juan county residing in the town of Silverton and 573 in the surrounding mountains the county's population was fairly evenly distributed between town and hinterland. But occupational, gender, and family distributions present a different picture. When we note that 61.5 percent of the county's men lived outside of Silverton and 76 percent of them were listed as miners whereas the 38.5 percent of males residing in the city accounted for the 65 percent of the county's merchants, we see once more that mining was the chief activity outside of town and commercial enterprise the business of Silverton. The county's sixteen women who lived outside the city and of whom only one was single were outnumbered by men at a ratio of 32:1. In Silverton, however, where the ratio of men to the eighty-eight women was close to 4:1, seventy-one women were married, thirteen were single, and four were widows. As we have already seen in the previous chapters, it was the wives—and particularly those married to the city's businessmen and entrepreneurs—who were responsible for the educational and cultural life that, as Nossaman has pointed out before, distinguished Silverton among other area communities. The single women were listed in the census either as the owners of or the boarders in a boarding house, as boarding with a family or living with relatives as sisters or sisters-in-law. Boarding house and boarding often were polite names for bordellos and prostitution, although such attribution cannot in every case be made with certainty. Of the four widows we are told that one was the mother of a four year old son, two earned their livelihoods as dressmakers, and one, aged twenty-three, was listed as a boarder.

And then there were the 124 children in the mining areas out in the county, seventy-three girls and fifty-one boys. Distance and terrain never allowed twenty-nine of them—eighteen of school age, twelve girls and six boys—to see the inside of Silverton's school house. Their education was limited to what they observed in their neighborhoods and perhaps learned from their parents. Of the other ninety-five children who in 1880 lived in town, forty-five were of school age, which, as stated in the last chapter, was set by law from six to twenty-one. Twenty-six of the forty-five were girls and nineteen boys. Compulsory schooling did not then exist, and not all the children were enrolled and attended the town's school.[3] According to the school's report at year's end, only twenty-two of the forty-five school-age youngsters had actually been present.[4] The statistics compiled for the state superintendent of public instruction, however, claimed thirty boys and twenty-eight girls for a total of fifty-eight as being of school age and thirty-two, just a little more than half, actually having been enrolled in Silverton's school. It appears that bureaucratic embellishment had not been an unknown art then either. But however that may be, the pupils' absence from school and their parents' lack of interest in their children's scholastic progress were to remain recurrent problems for Silverton's school throughout the rest of the century.

The people of Silverton and San Juan County were a polyglot lot. A little more than a third of them had crossed the ocean to make their way into the Rocky Mountain fastness. They came from many different European countries, with those from England, Germany, and Ireland leading the parade. The majority of those who had been born in the United States listed New York, Pennsylvania, Ohio, and Illinois as their state of birth. But with the exception of Leong Sing Lee, the Chinese laundryman, all of Silverton's inhabitants were white. There were in 1880 no Afro-Americans and no Native Americans living in town.[5]

As the 1880s began, the foundations for Silverton's rapid emergence as the commercial, cultural, and governmental hub of San Juan County had been laid. The first brick and stone buildings had gone up. One rose on Reese Street, another close to Cement Creek. In 1880 a restaurant opened its door on Greene Street and the Posey and Wingate commercial structure was completed.[6] By the year's end it hosted the Sunday school's annual Christmas Tree Festival on December 23. "Not less than 250 persons, including ladies, children and babies" were present, reported the *La Plata Miner*. "The program consisted of music, charades, tableaux, recitations...and in every respect showed good drilling, excellent taste, and successful accomplishment."[7]

Other events of similar nature were the literary sociables and spelling bees that brought together the town's white-collar families for an evening of entertainment and sociability. In January of 1882 the *San Juan Herald* reported on the literary society's "best meeting yet," a meeting that offered a potpourri of musical presentations by Miss Artie Earle, the town's beloved soprano, and the Honorable Harry B. Adsit, the county's clerk and recorder. This was followed by declamations, poetry readings, and a debate on the subject "Resolved that money has more influence over Man than Woman has." A month later a spelling bee also offered songs and the reading of an essay. The paper made sure that Silvertonians understood that a spelling bee—or, as the paper called it, a spelling school—did more than amuse. "It creates a taste for letters and inspires investigation into the peculiarities of the mother tongue, and, moreover, shows some of us the necessity for brushing up our musty orthography..."[8] Amusement with a touch of school lightened up the dreary winter nights.

Besides managing the literary amusements, Silverton's leading women supplied a support group for the Reverend Harlan Page Roberts, minister of Silverton's Congregational Church. Roberts had enlisted Mrs. Julia Earl, who had also spearheaded fund raising for the school house, her daughter Artie, the soprano, Mrs. Emeline Puckett and her daughters Belle and Huldah, the former teacher, now married, as well as the Hollingsworth daughters, Emma and Mary to raise funds for a church building he so desperately wanted. On August 20, 1880, he could witness the laying of the cornerstone and one year later, on the evening of Sunday, July 10, 1881, the dedication of the now completed First Congregational Church. By October a bell was installed in the tower and from then on called Silverton's Protestants to their Sunday services.[9] This accomplished, Roberts resigned from the ministry early in 1882 and served on the school board from December 1882 to September 1884. A Methodist Episcopal congregation inaugurated Sunday services in January of 1882.[10]

Other developments testified to the town's commercial and financial development. The San Juan County Bank began its business in July of 1880, and a safe arrived for its competitor, the Bank of Silverton, a few days later. In the next year telephone service from Lake City reached Silverton at the San Juan Bank.[11] A new weekly newspaper, the Republican leaning *San Juan Herald,* joined the *La Plata Miner* and published its first issue on June 30, 1881. As miners know, smelters are short-lived at high altitudes, and the town's Greene Smelter closed operations in October of 1880. It resumed its work under new ownership the next year at lower altitude in Durango.[12] As news of surprisingly large ore reserves north of town at the Red Mountain area arrived, Otto Mears in 1883–1884 added to his road network a twelve-mile stretch from Red Mountain to Silverton. His roads now connected Silverton with Ouray and allowed access to the prosperous Yankee Girl Mine from both directions. That road, wrote Helen Searcy, constituted Mear's "most spectacular accomplishment, and the one for which he is most noted.... In a setting of exciting hazards and unsurpassed grandeur and beauty it has never been excelled as a scenic route." Mear's final roadwork in the San Juans consisted of the reconstruction in 1883 of the Silverton-Howardsville-Eureka road and in 1886 of the four miles from Eureka to Animas Forks.[13]

In the meantime, a construction crew of the Denver and Rio Grande Railway had made its way up from the south via Durango and just missed Silverton's gala July Fourth celebration in 1882 by three miles and nine days. That, however, didn't stop the festivities that had begun on the third and continued on the fourth when the arrivals were picked up with horse-drawn wagons from where the train had dropped them.[14] Silverton now could look forward to convenient rail freight and passenger transportation to Denver, a much-acclaimed boon to the future prosperity of the town. By August, the railroad had installed a telegraph office in its Silverton depot. The appearance in the same year of four Chinese laundrymen relieved Leong Sing Lee of his lonely one-Oriental status and added a touch of cosmopolitan flavor to Silverton's otherwise European demographic make-up.[15]

For Silverton's leading families who prided themselves on their town's "religious, educational and civic facilities," the opening of saloons, dance halls, and houses of prostitution on the east side of Greene Street and on Blair Street was an eyesore and an insult to their sense of propriety. Alice Morris had sold her whorehouse on Blair Street to Alice Hanke in October 1880. At the same time the "Sage Hen" Jane Bowen and her husband William added to their Westminster Hall at Twelfth and Greene another structure at the corner of Twelfth and Blair. It served as both residence for them and their adopted eleven-year-old daughter and as dance hall and bordello. Across from it on Blair Street Samuel Thomas Stanton had built a one-story frame house that also served as a brothel.[16]

Silverton's cultural elite could only fulminate against these dens of iniquity and their denizens. The *San Juan Herald* thundered in its first issue on June 30, 1881:

> The brazen faced effrontery shown by some of the Amazons of this town as exhibited upon the public streets and thoroughfares, is becoming a nuisance to every decent and respectable citizen. There are crying evils in every community which it is difficult to suppress, but they should at least be kept within decent boundaries, and if countenanced at all, let it be under the strict surveillance of legal restraint.

Still, the "decent and respectable citizens" could take comfort in the knowledge that their school on Snowden Street was two to three city blocks to the west of the scene of Silverton's night-life. Presumably, their children, living with their parents on the westside of town, would not have occasion to venture east to Blair Street and, at night, would be kept safely at home. But who could be sure that, if not the children of the business elite, then those of mining families on the East side of town, would not explore their neighborhood? As Elliott West tells us, "Working class children panned for gold under sidewalks and in gutters in front of saloons and brothels, and as they neared their teens, boys were often invited in. Peddling bills and delivering goods, there was little they missed."[17]

The events that shook the town and brought home to everyone the perils that a wide-open nightlife would bring occurred on August 24, 1881, between twelve and one in the morning. Earlier in the evening Burt Wilkinson, Dyson Eskridge and some others, among them the "Copper Colored Kid," a gang of desperadoes, had ridden into town and burst into the Diamond Saloon. Located at the southeast corner of Eleventh and Greene, the saloon also housed Bronco Lou's brothel. Lou, however, was not present, because she had been jailed three days before for plying a miner with knockout drops and robbing him of his money.

While the Wilkinson gang lived it up at the Diamond, shooting out the lights in the process, the sheriff of La Plata County, Luke Hunter, arrived in town around eleven o'clock with warrants for the gang's arrest. He roused Silverton city marshal Clayton Ogsbury, and together they went to the Diamond Saloon. A shot rang out as they came close. Sheriff Ogsbury fell to the ground, mortally wounded by Burt Wilkinson. In the ensuing hunt for the desperadoes who fled on foot, the Copper Colored Kid was found and jailed. The next night a lynching party of Silvertonians broke into the county jail and hung the Kid from a beam in a nearby shed. The same fate awaited Wilkinson. He was arrested near Animas City and brought back to Silverton on Saturday, September 3. The next evening, Sunday night at ten o'clock, a lynching party appeared at the jail. Within a few minutes, Wilkinson was dead.[18]

Commercial development, literary ambitions, prostitution, gunfire, local politics, and, during the winter months, avalanche disasters provided the background against which Silverton's middle-class families nurtured their public school. During the winter session of 1879/1880 James Dyson had faithfully served his term as Silverton's teacher. At the end of the school year he handed over his duties to J. Homer Stewart, a miner, who had agreed to take on this assignment for the summer months. Stewart then was followed by Alice Dyson, James Dyson's wife. To the great relief of the school directors, Alice Dyson so successfully steered the school's fortunes during the winter term of 1880/1881 that the board of directors decided to extend school another month into April.[19] Mrs. Dyson had also seen to it that the "literary sociables" continued throughout the winter in the school house with various artistic and literary presentations. While she was absent for the summer, her place was taken by William K. Newcomb.[20]

When the school board met in September of 1881 it rehired Mrs. Dyson for a full year of three terms.[21] In October she could report that her students, divided into two classes, numbered twenty-two, amounting to about 41 percent of the district's fifty-four school-aged children.[22] Besides her teaching, Alice Dyson again hosted the "literary sociables" in her school that featured an exhibition of Longfellow's work and a temperance play. Her efforts were rewarded with praise from parents and pupils who presented her at the end of her term in June of 1882 with what the *La Plata Miner* described as "an elegant perfumery casket [*sic*] as a token of their respect and love."[23]

While teacher and students pursued their work in the school house, the school board—also called the board of directors— saw to the business of supporting and maintaining the school. In 1880 the Board consisted of its president, Dr. Robert H. Brown, a physician and former school administrator, John Rogers, the board's secretary, and the treasurer, Leander F. Hollingsworth, lawyer, and father of former teacher Emma Hollingsworth. On May 1, they published a letter in the *La Plata Miner* asking the town's taxpayers to support a special tax levy for the school to acquire new desks and seats for students and teachers, repair the ceiling, plaster the side walls, paint the inside woodwork, and point up the foundation.[24] When the board's new term began shortly thereafter, John Rogers was elected for a three-year term as president, and Dr. Robert H. Brown served for another year as secretary.[25]

In having been able to rely for teachers on the Dysons, both James and Alice, and on Homer Stewart and William Newcomb, the board could agree with State Superintendent Joseph C. Shattuck's remarks in his Second Biennial Report that because of immigration Colorado had not only been able to secure "excellent material for all vacancies," but also that a change was at hand. "The graduates of our own schools," wrote the superintendent, now need "the training of normal institutes, and this will be more urgent with each passing year."[26]

As school opened again on November 9, 1880 for the winter session, Alice Dyson and her students found the building ready for use, thoroughly repaired and refurbished. The *La Plata Miner*, however, seemed less convinced that all was in as good shape as it could be.

It commented:

> The district school began its sessions...but under how favorable
> auspices we have not been informed....The public school is a
> very important factor in any community, and should be no less
> so in our town. We trust therefore that the school in Silverton
> will be well patronized and successfully managed.[27]

In March the paper followed up its admonition with a suggestion
to the literary society that it charge admission to its events and
use the receipts to purchase desks and seats for the school.[28]
There is no indication that the Society acted upon the idea,
but two months later the board again ordered repairs as well as
desks and globes.[29] Thinking ahead to the need for expanding
into a new building the board also acquired three lots on the
school house block.[30]

In the fall of 1881, Dr. Robert H. Brown, the board's former
president and then secretary, had entered the election for county
superintendent of public instruction as the Democratic candidate. His Republican opponent was Cassius Marcus Frazier,
commonly referred to as C.M. Frazier, a lawyer who had served
as deputy county superintendent in Lee County, Iowa. As a
recent newcomer to Silverton, Frazier did not stand much of
a chance of winning. Nonetheless, the Republican *San Juan
Herald* thought his nomination to have been "a wise thing"
and declared him "eminently qualified." But when the voters
chose Dr. Brown by a margin of 252 over 119, the paper told its
readers that Dr. Brown was "a gentleman of culture and ability"
who had "enjoyed a large practical experience in the management of schools, both here and in the past....The people of
this county," the paper continued, "were especially fortunate
in being able to secure his valuable services..."[31] Mr. Frazier
was not pleased. He turned to the *San Juan Herald's* democratic competitor, the *La Plata Miner* to inform the children of
Silverton of his sad fate. This is what he wrote:

> My little children, who is this? I will tell you. It is a candidate. It
> is a defeated candidate. Why is he a defeated candidate? Because
> he was betrayed by his friends. They could not rule their party,

but they could ruin it. They were false friends, my little children, and very ungrateful, because they were mean enough to smite the hands that had heretofore served them....This candidate looks very sad, yet there is a smile upon his face. He is thinking of a day that will come when there will be weeping and wailing and gnashing of teeth among those who have betrayed the Republican party in San Juan county.

The *La Plata Miner*, on its part, happily acknowledged the Democratic victory and expressed its hope that "animosities will be buried, [and] sore heads bound up and healed."[32] Dr. Brown took over his new office as superintendent from his predecessor, Judge Henry O. Montague, in January of 1882 and was to serve in that post for the next four years.

The *San Juan Herald*, however, was not so easily mollified. Its publisher, George N. Raymond, charged the publisher of the city's competing paper, John R. Curry, "the nincompoop that edits the *La Plata Liar*," with commencing "a tirade of abuse" on the leaders of the Republican party. "If Curry had the brains of a two-year-old jackass," Raymond continued,

> he could see at a glance that the men he has opened fire upon are gentlemen of honor and principle (something that he cannot even lay the remotest claim to), but because they do not get down to the brute level and associate with him in his drunken carousals and help him fill his pockets with money with which he might satisfy his brute desires in the low houses of prostitution, he slanders them.[33]

Journalistic practice, it appears from this, differed little from the language of the San Juan mining camps and the city's saloons and bordellos.

CHAPTER 5

AT HIGH TIDE

As an evidence of the cost of doing business and living in Silverton previous to the completion of the rail-road, we are informed that all staple articles are now sold at from 40 to 50 per cent less than before the arrival of the road. Many articles are now sold for one-fourth of the former prices. While the prices of living and mining have been reduced, the price of getting ore to market has been reduced from $35 to $12 a ton.

—*La Plata Miner*, June 23, 1883

Silvertonians could well be tempted to number the years of their town's existence from the coming of the railroad in 1882. They expected 1883 to become a banner year. The *La Plata Miner* boasted in its December 30, 1882 issue that in the coming spring there would be in operation two smelters, two sampling works for crushing and sampling ore, and two concentrators. Half a year later it proudly listed as evidence of the town's success the existence of

> two hardware houses, 7 general stores, 2 clothing and gent's furnishing goods, 2 furniture stores, 2 harness shops, 2 boot and shoe shops, 2 flour and feed dealers, 4 millinery and ladies' furnishing goods, 4 meat markets, 3 drug stores, and 3 jewelry stores....In addition there are five hotels, 10 restaurants, 34 saloons, 5 blacksmith shops, 8 laundries, 1 auction store,

4 barber shops, 3 bath rooms, 6 tobacco, cigars, fruit and confectionaries; 4 livery stables, 2 bakeries, 1 theater, 3 dance halls, 1 photography gallery, 5 assay offices, 3 newspapers, 1 bank,[1] 4 doctors, 2 dentists, 3 mining offices, 9 mining engineers, 18 lawyers, and 294 dwellings.

Instead of two there were three sampling works, but only one instead of the expected two concentrators.[2]

The business boom was also reflected in the school censuses listed in the Reports of the State Superintendent of Education. Although there were fifty-four youngsters of school age counted in 1881 and seventy-two in 1882, that number had risen to 157 in 1883. Of these, so claimed the superintendent's report, thirty-six were enrolled in Silverton's school in 1881 and seventy-five in 1883. Alice Dyson, however, announced to have taught only twenty-two students in two classes in the fall of 1881. On the other hand, her successor as principal teacher, Miss Isabella Munroe—known also as Belle Munroe—together with her assistant, Miss Mary Gaines, reported in September 1883 that they had enrolled, though not necessarily taught, about seventy students in two departments of six grades each.[3]

On whichever of these figures we may rely, we note that San Juan County's school affairs were running in high gear. The new county superintendent of public schools, Dr. Robert H. Brown—known in town as "Dr. Bob"—played a commanding role in the city's social and cultural life. Besides having in the past served as president and secretary of the school board, he had also been active in 1881 promoting the establishment of a public library and was elected as chair of the newly organized literary society in November of that year.[4] When demands were heard in surrounding settlements for new schools to be erected, Dr. Brown was asked to assist the community of Animas Forks in setting up School District No. 2 for the sixteen school-aged children who lived in an area drained by the Animas River north of Burns Gulch. Brown warned them that aid was not going to be available until they had erected a school house of their own. This never came to pass, though a log house was leased and teachers were hired from the summer of 1882 to 1884.[5] It also was reported that another school district was to be organized at Howardsville to be in place by September 1883. But in the state's *Fifth Biennial Report of the Superintendent of Public Instruction* of 1887 San Juan County superintendent Dr. J.N. Pascoe reported that the county had only one school district, and that was No. 1 in Silverton.[6] It was not until ten years later that a public school began operating in Howardsville under the supervision of School District No. 1.

Silverton's school also was praised in the *Denver Republican's Annual* as being "always presided over by the very best collegiate ability" and having "a neat structure, well built and provided with all the latest modern improvements."[7] The board of directors of the Silverton District, however, aware of the rising tide of scholars that could no longer be accommodated in the old school house with its room limited to forty pupils, authorized the expenditure of cash to buy land. By June 1882 it had acquired seven lots for an expanded school house north of the old one.[8] In the fall of 1883 the primary grades under the tutelage of Mary Gaines moved into a rented building east of the old school house while the advanced students taught by Belle Munroe stayed in the old house. The *La Plata Miner* expressed itself well pleased with these developments. "The teachers," the paper wrote, "are entirely competent and...the board is composed of men who labor for the permanent advancement of the educational cause in our town."[9]

To the board members it had become obvious that the time had come now to ask the voters for funds to build a new school. Their only question was whether the money should be raised through taxes or through bonds. At meetings in May of 1884 taxpayers opted for a $10,000 bond issue to construct a four-room, two-level school building with a capacity for 400 or more students. While nobody expected that many students in the next few years, the building would be ready in the future when the need for a high school was expected to arise.[10] Construction began in September 1884, and teachers and students moved in the following January. Of Silverton's nearly 100 children, reported the *La Plata Miner,* 80 percent were enrolled.[11] As in the past actual enrollment and attendance figures, however, were considerably smaller. Few, if any of the nineteen to twenty-one age-group would ever avail themselves of their opportunity to go to school at their age. The *Silverton Democrat*, always ready to counter a *Miner's* encouraging statement with disparaging remarks, complained about the school board's decision to hire a building superintendent. "It is rumored," wrote the paper, "that incompetent church favorites as building superintendents are not relished by tax-payers."[12] The target of the remark, it appears, were Silverton's school and church supporting ladies.

The new school building measured thirty-eight by sixty-eight feet and was solidly constructed of stone and timbers. It featured a basement and two floors with spacious hallways running lengthwise through the building. The hallways were connected by a staircase that lead from a centrally located entrance hall to the basement below and the floors above. Furnace and storerooms filled the basement, and classrooms, cloakrooms, and the principal's office were located on the floors above at both sides of the central hallways. The building was introduced to the general public on the evening of March 5, 1885 with a community spelling bee, sponsored by the Congregational Church. Former teacher Alice Dyson served as referee. As the *Silverton Democrat* reported, the contest began with fourteen people on each side and ended when only five ladies and one gentleman were left. The word on which everyone lost was "Ophicleide." Unfortunately, the paper did not—or could not—explain what it meant.[13]

School affairs were now in the hands of James Dyson as board president. Dyson served in that capacity from May 1883 to August 1885 when he was replaced by attorney Nathaniel E. Slaymaker. C.M. Frazier, who appeared to have overcome his disappointment over his loss in the November 1881 election for county superintendent, had been elected treasurer of the school board in the May 1884 election and served in that capacity until 1887. His name showed up again in 1888 when he held the post of school board secretary for another two years. Harlan Page Roberts, the former minister, had been the board's secretary since December 1882, but left town in September 1884 to pursue a legal career in Minnesota. His temporary successor, John Montgomery, was appointed by Superintendent Dr. Brown. Montgomery's place on the board was hotly disputed in the May 11, 1885, election in which Jonathan W. Fleming beat

James H. Robin, a well known businessman and clerk of the district court, by a count of 129 over fifty-nine votes. As the *Silverton Democrat* noted, "a considerable amount of money was staked on the result.... Quite a number of ladies voted," and Fleming drew most of his support from "the upper part of town."[14] James H. Robin had been no match for the voting power of Silverton's "cultured ladies."

Dr. Robert H. Brown, who had defeated Frazier in 1881, was reelected as county superintendent in November 1883 during a massive voter turnout. He defeated school board president James Dyson by a vote of 687 to 505. It is not clear whether Brown, a Democrat, retained his office because of his party's victory or because the voters wanted to keep Dyson as school board president. Both Dyson and Brown were capable and admired men, and the voters saw to it that each stayed in his post that he served so well. Brown continued to serve for another two years as superintendent until he was succeeded in January 1886 by the town's physician, Dr. J.N. Pascoe.[15]

As in years before, the board's continuing problem had been to secure competent teachers for their school. Alice Dyson, her husband James, and Isabella Munroe had been teachers who had served exceptionally well, and Miss Munroe was to stay as principal teacher until 1885. Her assistant, Mary Gaines, left in June 1884 and was replaced by Miss Armes as teacher of the primary department. Miss Armes' appointment did not please the *Silverton Democrat* whose editor wondered why the board had to import a teacher "when we had competent material among our own residents."[16] The *Democrat's* Republican competitor, the *San Juan Herald*, then sought to even the scales by commenting that it was "one of the prettiest sights of Silverton...to see the handsome petite schoolmarm of the primary school surrounded by her innocent cherubs on their way down Reese street to the school."[17] When Belle Munroe announced her retirement in March 1885 the petite Miss Armes took over as principal and another out-of-towner, Cora Downer, was asked to teach the primary department. By July, however, Miss Armes had left town, and Belle Munroe was asked to step back in as principal until another teacher could be found. When school opened in September Cora Downer advanced to the principal's place and Frances Tracy began her assignment as assistant teacher.[18]

Young people, too, had made the columns of the *La Plata Miner* during these years. The paper complained in March 1883 that

> certain young people of this town, of both sexes, are in the habit of throwing chunks of coal against doors to frighten the inmates, dancing the racquet on the streets, and, to cap the climax,—visiting young men who are afflicted with the measles.

The paper approved of the visits to the sick, but, it added, "for mercy's sake, no more 'racket.'" In June it editorialized about the young ladies who were "getting more plentiful in Silverton than cranberries in a cranberry marsh" and could be seen "every evening in groups of three or more..." It advised the men on Greene Street who were listening to music or talking about their "prospects" to "catch on." Shifting its attention from the girls who "danced the racquet" and populated the "cranberry marsh" to those who were taught in fashionable schools and universities, it complained that these were not taught

> the art of home-making and home-ruling...[and did] not know how to make a loaf of bread or to set a table, or to iron a napkin, or to make a bed becomingly. Is it expected that servants shall do these things?

The writer answered his own question by stating that it would not harm anybody "for the mistress of a household to know how to calculate an eclipse, but it is dangerous for her to be eclipsed by her Bridget."[19] Was the *Miner*, too, casting aspersions against the town's ruling church and literary ladies?

On the subject of young ladies the *San Juan Herald* had to add some observations of its own. It did not think that "young ladies were getting more plentiful than cranberries in a cranberry marsh." To the contrary, its editor thought that there were far too few. "There are thousands of young men in the San Juan, today," he wrote, "just dying to get within the reach of a puritan beauty..." Providence should send several car loads of them. New England girls, he assured his readers, "will find homes here and will go as readily as water-melons just now." When a reader wrote back that "a good, nice, healthy girl, who can make her dress and get up a good supper for company, and is not ashamed to wait on the table while they are eating, is just worth about $1,000 a pound," the editor added: "If they can get out prospecting and jump claims they ought to be worth $2,500 a pound, providing they don't lose any weight on a prospecting tour."[20]

Regardless of what the papers would write or insinuate about them, Silverton's ladies were satisfied with the city's cultural and civic activities during the summer months of 1883. The Congregational Church played host to an Episcopal service that included "a double quartette, composed of the best musical talent in Silverton." A few days later Robinson's circus arrived which the *La Plata Miner* saw as "evidence of our importance," and praised for its excellent performance and the "good order maintained—no fights, no drunks."[21] The Fourth of July brought what was to become an annual event, still continued in the twenty-first century: Truck and ladder races and water fights between Silverton's and Durango's fire companies. In 1883 Silverton won both of them. And there were horse races, picnics, excursions, and "other innocent amusements." The day ended with a ball at the newly finished brick Grand Imperial Hotel that led the *Miner* to comment:

> People throughout the East think that Silverton is almost outside of the limits of civilization but if they could have looked into the ball room and seen the elegantly dressed ladies and we will add good looking ladies, and the delicate attention shown them by our gallant young gentlemen, they would probably conclude that we were as civilized and refined as the majority of the people in the older states.[22]

And when school started on September 17 the paper fairly gushed with pride and satisfaction. Belle Munroe reported she was ready to teach algebra, physiology, philosophy, and other subjects to more students than ever had been enrolled in the advanced department. Members of the school board, wrote the paper "have shown excellent judgment in selecting teachers . . . [and] furnishing of the building."[23] Congratulations were in order.

But not everything smelled like roses in Silverton in the summer of 1883. The *San Juan Herald* had taken a tour of Silverton's alleys and side streets and found them to be "in the worst condition imaginable." Filth and stench permeated the air and flies were everywhere. By September the town's physician Dr. James Brown had reported over 100 cases of dysentery. The *Herald* called it "a wonder that our people are not all sick."[24] When in June proceedings were initiated in district court against several gamblers, Denver papers telegraphed that Silverton "was in the hands of a mob, who were not only expected to burn down the town, but to rob the bank and people indiscriminately." This misstatement, the *Miner* reported, was the work of "parties, not interested in San Juan," who represented "malicious devils who sent forth . . . damnable lies." But, the *Miner* assured its readers, "a man is as safe here as in any town east of the Mississippi River."[25]

Matters were not helped much, when two weeks later the *Miner* had to admit that "there is an opium den in town with a good line of patrons, both men and women. Some of them are of the better class." Came November, the issue had not come to rest. "Several Celestials" the paper wrote, "paid heavy fines." But that wasn't enough. "Let the good work go on. Arrest every occupant of a building where opium is found, and keep an eagle eye on every Chinaman in town."[26] The paper did not say why that eagle eye should not also have been kept on "some of the better class."

Both papers evidently regarded it as one of their duties to keep Silvertonians informed of what their non-Anglo neighbors were doing, and neither of them was very circumspect in its choice of language. The *Herald* reported in May of 1882 the arrival of "a few more of the heathen Chinamen, as also several gents of African persuasion." It then gave a lengthy disparaging description of "two inebriated darkies" who "were parading the streets one evening this week closely locked arm in arm. . . . John Chinaman," it added in July, "is well represented in Silverton, and the result is three or four Celestial laundries . . ." In October 1883 it reported the killing of Ju, an Apache chief.

"His demise," it added, "is certainly a cause for rejoicing, as Ju was a bad Injun."[27]

But it wasn't only the Celestials and Apaches in another county about whom the good citizens of Silverton learned in their papers. They were more concerned about their immediate neighbors, the Southern Utes, and they took notice of a petition sent in February 1884 to the U.S. Senate and House of Representatives by prominent business men of Durango. In that petition, which the *La Plata Miner* headlined *Bounce the Utes*, the Durangoans objected to "the continual quartering of a useless, idle, predatory and dangerous band of savages...on a strip of land known as the Southern Ute reservation." That land, asked the petitioners, should be released "to its proper and God appointed ends," and the Utes should be removed "as speedily as possible and their said reservation opened up to settlement..."[28]

There is little evidence, however, that Silvertonians then paid much attention to that call for Indian removal. They had other and more immediate problems to worry about. Beginning in February 1884 snow kept falling and would not stop. The Denver and Rio Grande, their lifeline to the outside world, was blocked from February 4 to April 17 by snow and avalanches. The mines closed and miners and laborers came down from their sites and crowded into town. Food ran short, and as an eyewitness from Chattanooga reported,

> the snow was 12 feet deep...[that] we could no longer shovel the trail; so we dug a tunnel under the snow from our cabin to the barn in order to take care of the stock....[A]ll fresh food was gone and practically nothing left but flour. Starvation was in sight for people and stock animals alike.

The local hero was Sam Herr who, with some others, went down the Animas Canyon to repair the broken telegraph wire. Herr also braved the two-day trip over the forty-five miles to Durango along the railroad track more than once "with the U.S. mail on his back and snow shoes on his feet." As the *La Plata Miner* commented, "it looks like a reckless venture on Mr. Herr's part and is certainly an unselfish one. *Dieu vous garde.*"[29]

Figure 5.1   Snow Tunnel Near Silverton

Cut off from the world and beleaguered by avalanches, falling snow, and declining food reserves, Silvertonians kept up their spirits as best as they could. Belle Munroe managed the school and reported fifty-eight days of teaching thirty-seven enrolled children in the grammar department during the winter term ending March 28.[30] Adult society fell back on parties—parties, that is, of rather different characters.

In February the town's soiled doves set the pace at the *Maison du Joie*. It was attended, reported the *Miner*, "by the *elite* of the *demi monde*." Elite or not, blows were thrown and Dora Taylor, one of the madams, "engaged in a hair-pulling and black-eye circus." Dora was arrested but released when she promised "to go and sin no more." Apparently she did not keep her promise because, as the paper reported a month later, she "and several of her soiled doves left for Durango by the D.& R.G. route, on foot..." Whether they, like Sam Herr, traveled on snowshoes or stomped without them through the snow the paper did not say. However they did it, it was a grueling two-day trip. Still, it may have been preferable to the one chosen by their colleague Mabel Coulter who, the paper wrote, "attempted to go by the chloroform route to another kind of place," though, the paper added, "[she] did not get through."[31]

The 1884 soiled dove exodus, however, did not produce lasting results. By the fall of the next year, the *Silverton Democrat* raised its voice in disgust:

> It is not a credit to the town, but a fact nevertheless, that nearly 130 prostitutes find a living here. About nine tenth of them support 'tin horn' gamblers, in addition to supporting themselves.[32]

The town's tax on prostitution was too tempting a source of income for the community to put an end to the business.

A month after the hair-pulling circus on Blair Street a "social Thursday evening" was held at the residence of the Dysons. The high point of the affair was the reading by Isabella Munroe of her article on the snow blockade, which, so wrote the *Miner*, "was full of telling hits and established for that lady a reputation for keen satire and sparkling wit." Besides Mr. Dyson as school board president, Mrs. Dyson as former teacher and Miss Munroe as current principal, the party brought together a cross-section of the town's social elite, including the owner of the *Miner*, John Curry and his wife, and John Montgomery who was soon to be appointed to a vacancy on the school board.[33]

By mid-April, Silverton had survived its period of isolation. "Last evening at 6 o'clock," reported the *San Juan Herald* on April 17, "a passenger train from Durango arrived in Silverton...Yesterday a large freight train containing coal, merchandise, etc. came in, the canyon having been opened yesterday by a force of workmen numbering 250. The track from here to Durango is now clear..."[34] Silverton's citizens breathed a sigh of relief. They now could look forward confidently to the continuation of prosperous times during the remaining years of the 1880s and the approaching 1890s.

CHAPTER 6

SCHOOL AND TOWN IN THE 1880s

Papers for the People! One an alleged Republican paper run by a Democrat, and the other a Democratic paper run by a Republican; and then to think that they will call each other boodlers.

Thus wrote the Republican *San Juan* on September 15, 1887. It lampooned John Curry, who had been the founding editor of the *La Plata Miner* and now edited the *San Juans* Democratic rival, the *Silverton Democrat,* by citing Curry's battle cry, a paper for the people: "It is, John," the *San Juan* wrote, "but not for them to read." As for the alleged Republican paper run by a Democrat, that was the *San Juan* itself. Its proprietor and editor, George N. Raymond, had just given a one-half interest to Charles Day. A few months later the two would pass on the editorship to Charles W. Snowden who intended to run the paper as a Democratic sheet. Silverton's newspaper scene had been fluid all along, and the rivalry of the early 1880s between the supposedly Democratic *La Plata Miner* and the Republican *San Juan Herald* had now been replaced by that of the originally Republican but now soon to be Democratic *San Juan* and the *Silverton Democrat.*

Yet no matter how Silverton's newspapers' partisan loyalties twisted and shifted, they all were steadfast boosters of their community. Only fifteen months earlier in June of 1886 the *La Plata Miner* had still proudly reminded its readers that it had been in existence for twelve years without missing a single issue.

It recounted Silverton's growth from a small hamlet and mining camp in 1874 into a prosperous city that today in a single year shipped a million and half dollars worth of silver ore.[1] The *Miner* expressed well the dominant sentiment of Silverton's citizens. There had been a time, a decade or so ago, when Silverton had been nothing more than a mining camp and hamlet, yet its residents had loved to speak of it as the metropolis of the San Juan. Now it had in fact become a prosperous city, a metropolis of sorts, and when the *Miner* called it a "camp," this was no longer a slight, but an expression of pride and honor. Look, it said, how far we have come! We need not hide our roots; we can be proud of them!

Two years later, just before it passed into the hands of Charles Snowden, the *San Juan* felt it appropriate to paint a similar proud picture of "the gem of the Rockies" and "the hub of the mining world." It told how its permanent population of 1,500 people easily doubled during the summer when many of the miners returned and other residents came back to enjoy Silverton as their summer resort. A "vast amount of work and activity" ensured the economic welfare of the city, and its families helped "disrobing herself of the garments of hurrah days and dressing in the garbs of civilization." There were three hotels, six restaurants, and thirteen saloons, the paper recounted, the latter offering the best brands of wine, liquors, and cigars and billiard games as well. The *San Juan* did not fail to mention Westminster Hall, the dance hall of the famed Jane Bowen and her husband William, but did not reveal whatever unsavory activities might have been going on within its walls. Silverton's Italians found a welcome in Louie Regalias saloon, and Mr. Adolphus "deals out the beer to the German residents of our city." On the "civilized" side of the ledger, the paper listed the two churches, the Congregational and the Catholic, and described the school house as "one of Silverton's most notable structures." It mentioned principal and teacher as well as the members of the board of directors. It did not fail to refer to city hall, county jail, fire department, two business blocks, the First National Bank,[2] the two newspapers, the drugstores, meat markets, and a long list of city officials and individual businessmen. It listed the city's secret societies, commenting "in this respect Silverton excels."[3]

Such pride and self-esteem demanded obligations that the city owed its less favored citizens and the hundreds of hard-working miners—the "boys," as the *Miner* called them—who worked in the mountains above and around Baker's Park. They came down every so often to town after an arduous day of travail and danger in the mines. Some would have taken their meal in a boarding house at the mine; some would have done so in town. But after that, nine times out of ten, wrote the *Miner*, they would end up in a saloon. There it would "not take a great while to make an ordinary habituate of these places a drunkard in spite of himself. This state of affairs should not be allowed to exist in this growing and prosperous town, even though it be a mining camp." The *Miner* suggested in 1886 that Silverton's citizens come to the aid of the Knights of Labor and help them finance the establishment of a reading room or library where miners could sit and read magazines, newspapers and books and write letters home. "Give the boys a reading room," pleaded the *Miner*.[4]

A week later, twenty men and twenty women met at the school house to take the matter under advisement. Nathaniel Ellmaker Slaymaker, the president of the school board, chaired the meeting that founded an organization that would over the next few months establish the desired reading room and library. Most of its members were Silverton's civic-minded ladies, Mrs. John (Maggie Williamson) Montgomery, the future school board president and Mary Hollingsworth, the town's new postmistress, among them. They were joined by a few gentlemen, including Dr. J.N. Pascoe, the current county superintendent of public schools, and James E. Dyson, former teacher and school board president. They could celebrate the successful completion of their project with the opening of the Public Library on June 8 over Kruschke's store on Greene Street. Three months later the librarian, Mr. E. Robin, reported that 1,346 visitors had registered at the library, and most of them, alas, had been residents of the city, not "the boys" from the mines.[5] By that time, Mr. Slaymaker had been elected Silverton's mayor and Mrs. Montgomery had taken his place as school board president.

While proper Silvertonians thus furthered the cause of enlightenment and literacy on Greene Street, their demimonde sisters on Blair Street upset the members of the toboggan club when they took over their slide one evening in February. The *Miner* suggested that with the help of a town ordinance "these frail women" be confined "to their own quarters...and fined or imprisoned if found in the respectable portions of the city....A lady can hardly walk a block on any street in town without some tipsy woman running against her."[6]

Jane Bowen and her saloon made the paper at least twice in March. There had been "a grand soiree (?)...to celebrate the return of her worthy spouse" and the festivities had "passed off without so much as a riot," reported the *Miner* on March 13. Two weeks later, however, there had been a fracas when Dutch Lena had "snatched a beer glass and struck Irish Nell on the face, cutting a fearful gash on her lower lip." Lena was arrested and locked up and, later in May, was let go again. The grand jury had failed to indict her. By that time, the paper announced, Blair Street's population had received its seasonal influx of colleagues and competitors as "the sisters of doubtful morality" had "come to town for the summers campaign."[7]

As the 1880 census had revealed, the largest non-English speaking group of immigrants in Silverton and San Juan County were the Germans. They brought with them and continued to practice their language and their customs. On March 9, 1886, they celebrated the thirty-ninth anniversary of the founding of the American Germania order. The hall in which the occasion took place was decorated with a banner that read *Freundschaft, Liebe, Humanitat.* "The members, about thirty in number," wrote the *Miner,* "marched into the hall from the anteroom, in full regalia, and after circling around the room, took their seats....A beautiful hymn was sung, and the orations commenced. Everyone used the German language..."[8] From the description given by the paper it appears that among the participants residents of Silverton and their friends outnumbered the "boys" from the surrounding camps.

In the municipal spring elections of April 1886 the *La Plata Miner* supported the outgoing school board president, Nathaniel E. Slaymaker, who had been nominated for mayor by a Peoples Convention. The *Silverton Democrat,* which had been bought early in February by John Curry, promoted a ticket headed by Dempsey Reese for mayor and Curry for treasurer. According to the *Miner*, the Silverton Democrats were "thoroughly disgusted" to see the editor of their pet paper running for office. Their town committee decided not to nominate a Democratic ticket at all. They favored, so the *Miner* wrote, the election of "good men."[9] Apparently not living up to the Democratic town committee's standard of "good men," the *Silverton Democrat*'s Reese-Curry ticket went down to defeat, and Nathaniel E. Slaymaker exchanged his hat as school board president for that of mayor.

During the months of 1886 and 1887 the town's mood seemed placid, and for some rather a bit too placid. The *San Juan* had noted in the fall of 1886 that for months there had not been any complaints about the behavior of Silverton's young people, and that was unusual. The paper even urged them at Thanksgiving to make themselves heard and "get up a dance," because "this town is getting altogether too dull in a society way." We don't know whether this appeal had brought any results, but it had been heard by the town's young teacher, Susie Robin. On April 8 of the following spring she and her sister Isabella "Belle" threw "a very enjoyable party" at their residence to celebrate Susie's reappointment for the spring term.[10]

By November 1886 Silvertonians thought it a certainty that Otto Mears would build the long anticipated Silverton Railroad at least up to Chattanooga and probably to Red Mountain. When in December telephone lines between Silverton and Red Mountain were being installed, it seemed a certainty that after the snow had melted, railroad construction would begin. And so it indeed happened in June of 1887 when Otto Mears gave the signal to start. In constructing his railroad, he utilized the grade of his Rainbow Route toll road and completed the work to Red Mountain by September 1889. Two months later the road was extended to Ironton.[11] Few doubted that with its completion Silverton's importance as shipping point for the ore would reach new heights.

Reports also reached Silvertonians about unrest in the ranks of American labor. They read in their papers about the growing strength of the Knights of Labor and their strike against the railroads of the Southwest. Closer to home they heard about the coal miner's strike in Boulder and Weld counties. The *San Juan* expressed sympathy for the miners who had asked for a return to their former higher wages at a time when the mine owners had shown no willingness to lower the price of coal for the consumer. Silvertonians had mixed feelings on the issue. Being well aware that they were beneficiaries of the hard labor miners carried out in their vicinity, they were sympathetic to their cause, as they had shown in their campaign for the city reading room and library. But they were made uneasy by the talk of work stoppages and strikes, and by the unexpected rise in national popularity of reformers such as the single-tax advocate Henry George. Mr. George, they read in their paper, had easily outpolled the Republican candidate in the 1886 New York mayority race. Having garnered 75 percent as many votes as the Democratic winner he had shaken the city's Democratic establishment as well.[12] Silvertonians could only wonder what these events across the country might ultimately mean for the economic and political stability of their town.

The winter of 1886/1887 then brought the usual interruptions of daily life with its unpredictable invasions of snowfalls and avalanches. On January 14, 1887, one of these had blocked the railroad track and forced the train headed for Durango to stop four miles out of the Silverton depot. The engineer uncoupled the engine from the train to steam ahead and shove the snow off the tracks. While he was thus occupied, another avalanche descended on the unhooked express mail car and passenger coach and pushed them in the Animas River. As it was, the train's only passenger had left the coach and had walked after the engine to watch it clear the track. The express agent in the mail car saw what was coming and jumped out of the car in time. So there were no injuries, and the engine with a box-car, agent, and passenger safely reached Durango. A work train came up the canyon the next morning, repaired the track, and put the mail and passenger cars back on the rails. For Silvertonians, used to the seventy-two day avalanche blockade of 1884, this was nothing to get excited about.[13]

As April arrived, Silvertonians and miners who had spent their winter months in warmer climes returned to town. The next month, the *San Juan* reported, "a lady and a gentleman" increased the town's colored population. By June, however, most of the town's colored people had left Silverton with the circus. "At least none could be found yesterday to do any work," wrote the paper on June 30. But "a forlorn-looking Chinaman" showed up "in search of a location for a washee house." The town's soiled doves also made news again. "A fair courtesan" was arrested on May 21 "for parading the streets on horseback and being intoxicated." After being fined in court on the following Monday, she left by train, waving "farewell to the parties who would not let her run the town." A couple of weeks later, Lizzie Fischer was arrested "for promenading between Reese and Greene streets" and was also fined. The paper, however, did not tell whether she, too, left for more hospitable locations.[14]

In August 1887 reports of Indians on the warpath alarmed the city, and a meeting was called, as the *San Juan* put it, of "the Silverton Boys Anxious to Join in the Skirmish." Those eager to shoulder their muskets had proposed to ignore the general government "that is composed by philanthropists of the east" and to put an end "to the yearly butchery perpetrated by the dusky red skins." About thirty men thus motivated assembled in city hall, offering their services and specifying in which capacity they wanted to serve. They formed a band of their own and pledged their readiness to follow the call of their elected captain, William Keith, whenever roving bands of Utes were known to approach the city. Among the volunteers, for the most part young men, was one of their elders, none other than Superintendent Dr. J.N. Pascoe, who "offered his services as surgeon-general with no salary." A certain J.W. Cory, who proposed to go along as chaplain, was told by "the boys" that "no one but good fighters were wanted."[15]

While these local events made daily headlines, Silverton's newspapers paid occasional attention also to national events in education. The *Silverton Democrat* printed an article first published in the *Journal of Education*. The *Journal* had proclaimed the public school to be "a vital element in American life" that had now become indispensable. More and more Americans had left their churches, amusements had taken the place of lectures, and the popular press competed "for sensations sensationally displayed." It was the public school's mission to step in and respond as it can and would to these deficiencies in American life.[16]

The *San Juan* closed out the year with a lengthy reflection on the language issue that had come to the fore in many communities across the country: should public schools be forced by state law to offer their instruction in basic academic fields in the English language only, or could other languages be taught or used in districts where local populations so desired? A case had arisen in Saint Louis where, under the guidance of a school board heavily influenced by the city's German community, German had been taught as a regular school subject in the public schools. In an 1887 school board election Democratic opponents managed to change the composition of the hitherto Republican-leaning board, and the new board terminated the teaching of German in the schools. The *San Juan* approved:

> We cannot help but believe that they [the German people in St. Louis] are radically wrong...The English language is the exclusive language in the United States in all public educational matters. Other languages are taught in our universities and various great institutions of learning, but the people of St. Louis will do an unwise act if they sanction other than the mother language to be taught in our public schools.[17]

There is no evidence to show that the language issue had actually arisen in the Silverton school, but, as we shall see, national debates on public education were beginning to penetrate even the mountain fastness of the San Juans and Silverton's people would become participants.

In the spring of 1888 the *San Juan* spoke up for Senator Teller's bill to make education compulsory for American Indians and predicted, that if the bill were passed, it would mark "another important step in the progress of civilization." A few weeks later it published a letter to the editor which praised Senator Henry W. Blairs 1888 bill to combat illiteracy in the nation through a program of federal aid to public education that would distribute funds equally to black and white schools. The writer advocated compulsory education on a nation-wide scale.[18]

On the local education scene May 3, 1886 had been a red-letter day. Women and mothers interested in school affairs could proudly point to Silverton's first elected woman president of its school board. Mrs. John (Maggie Williamson) Montgomery received twenty out of thirty-one votes cast and was to serve for the coming year together with Jonathan W. Fleming as the boards secretary and C.M. Frazier as its treasurer. After school closed for the summer vacation on June 4, the board immediately began searching for new teachers. Cora Downer, last year's principal had left Silverton for her home in Iowa, and though there were hopes expressed in the *Miner* that she might return in the fall, the board on August 4 hired Mary Bertha "Mollie" Stockman (Mrs. Emerson Warren) Hodges and Susie Robin, sister or half-sister of James H. Robin, as assistant teacher for a three months term.[19]

When the fall term began on September 11, Mrs. Hodges was designated as principal teacher with supervisory authority over all departments. Salaries were paid at the same rate as during the previous year, $85 per month for the principal teacher and $75 per month for the assistant teacher. The board also agreed to let the German Society rent the school's North Room for its activities. When the fall term ended in December 1886, the board extended the teachers' contracts to the end of the school year in June 1887.[20] The town's newspapers, the *La Plata Miner,* the *Silverton Democrat,* and the *San Juan,* a Republican paper that had appeared for the first time on October 14, all published the school reports which showed that by the end of 1886 seventy-three pupils had been enrolled in Silverton's school. Forty-four of them had attended the primary department, thirty-four in the B grade and ten in the A grade. The older ones numbering twenty-nine had studied in what was called the advanced room, sixteen in the B grade and thirteen in the A grade. When the school year ended in May 1887 the total had risen to eighty-two, forty-six of them in the primary department and thirty-six in the advanced room. Besides giving these statistics, the reports also listed the names of each pupil, their average scholarship and deportment ranking, and the number of days each one had been in attendance. Privacy concerns obviously played no role in these revelations!

During the school year 1886/1887 both county and state superintendents of public instruction, Dr. J.N. Pascoe of Silverton and Leonidas S. Cornell in Denver, were Republicans, and both were received well in Silverton when they appeared as speakers. Dr. Pascoe, one of the town's two physicians, was well known to Silverton families and often spoke to the school children and adult audiences on his favorite topic of physiology. It wasn't always clear, however, that his young audiences really understood what he was saying. Once, after he had addressed the Silverton pupils, "a young miss of eight," reported the *San Juan*, "went home and told her mama that she had had a fishy-colicky lesson." Leonidas S. Cornell came to Silverton occasionally and spoke about the importance of public education and the need for communities to support their schools enthusiastically and generously. He was, so wrote the *San Juan Democrat*, "a pleasant speaker and never fails to interest his listeners." Another speaker who touched on educational topics that were of interest to Silvertonians was Professor Hale of Central City. He addressed the annual meeting of the State Teachers' Association in Colorado Springs in December 1886 and asked whether secondary education should be the same for boys and girls.[21] That was a question that within another ten years would excite Silvertonians when they debated the opening of a high school of their own.

As I have noted, the winter and spring terms of 1887 had brought a gradual but steady increase in the number of children enrolled in the school. Both the *San Juan* and the *Silverton Democrat* faithfully reported events that took place in the school house, such as the spelling bee that was to take place on February 8. "No one should remain away," wrote the *San Juan*, "even if they do not wish to spell, as the spectators will undoubtedly be ones who will have the most fun." Both papers also enthusiastically reviewed the entertainment program that the schoolchildren presented in March under the direction of their teachers, Mrs. Hodges and Miss Robin. The Silverton Cornet Band provided the music for the occasion and Superintendent Dr. Pascoe presented another lecture on physiology. In it, according to the *San Juan*, he "satirized the prejudices and superstitions of the present day..." The school's winter term ended on March 25 with literary exercises where the children sang songs, presented recitations and declamations, and showed their hand-crafted art work. Mrs. Hodges and Miss Robin were acclaimed by everyone as outstanding teachers and the board reappointed them for the rest of the school year.[22]

The spring term shifted the spotlight from the teachers and pupils to the members of the board. The question of an extra month of school in town was raised, but the board declined to consider it. Sufficient funds were not available, and the school house had to be plastered and painted. "It is claimed that the school building is becoming quite weak in some of its parts," the *San Juan* was to comment on June 16, "[and]…it is better to fix it now than to wait till some of the little children are hurt." But board members thought a three months summer school at Red Mountain feasible because there were between thirty and forty children of school age in the district, many of whom were willing to attend. As it turned out, sixteen pupils showed up in late April to be instructed by J.B. Perry, a man not otherwise identified. The school did not really get underway until May 30 when Miss Kittie Real took command of it. Miss Real and her sister Lena Real had recently come from Durango and had opened a millinery and dress making shop on Silverton's Thirteenth Street. Kittie Real then happily accepted the position of teacher of the Red Mountain Summer School.[23] She carried out her task successfully, teaching her more than twenty children. The school closed as planned on August 20.

While in operation, the school and the idea of running a school during the summer months became a favorite topic for the *San Juan*. "We often wonder," the paper wrote, "why our schools are not taught in the pleasant season instead of during the stormy months of winter? There are about two months when it is almost impossible for the little folks and teachers to reach the school building." The people of Red Mountain ought to be congratulated for their good sense, the paper concluded.[24] The *San Juan's* comments illustrate well how living conditions in the mining economies of mountainous areas differed from those in agricultural regions elsewhere. Mining operations, though shut down at times during the winter months, continued for the most part year round, and children were used for labor in the mines only to a limited degree. Agricultural labor, however, was seasonal, and children were put to work in the fields during the summer and were not available for schooling.

May was also the time for the annual election of school directors. C.M. Frazier's three-year term as treasurer was up, and though there was some talk of reelecting him, a group supporting Dr. James W. Brown, the county coroner, and another backing David Ramsey made it appear as though a lively campaign was about to break out. But Dr. Brown declined to run, and David Ramsey was elected unanimously. But then Jonathan W. Fleming, the board's secretary, resigned his seat abruptly, and Dr. Pascoe, the county superintendent, had to step in and appoint a successor. His choice fell on coal merchant W.J. Forsyth, which the *San Juan* acclaimed as a first-class one.[25]

As the school year of 1887/1888 began the new board of directors, apparently not satisfied with Mrs. Hodges and searching for a male principal, hired Professor C.M. Kiggins of Kirkville, Missouri, for a probationary period of three months. Kiggins was to be paid $110 a month and was to teach the advanced department. Susie Robin was rehired as teacher of the primary department at $75 a month. If they were to succeed in raising or receiving from the State Superintendent of Public Instruction sufficient funds, the board members agreed, they would employ Mrs. Hodges on the same terms. This, however, did not happen. At the board's October 31 meeting, the minutes read, there was not "sufficient consent to employ another teacher [and] it was deemed best to only have two teachers." At the next meeting on November 13 the board reaffirmed Mr. Kiggins' appointment as principal and raised his salary to $133 a month. It also greeted its newest member, the town's druggist Benjamin Austin Taft, who took over the post of treasurer from David Ramsey who had recently resigned.[26]

As the fall election for County Superintendent of Schools approached, Republican Dr. J.N. Pascoe, ready to join the Indian fighters as surgeon-general, did not run. His place was taken by James E. Dyson, the city's engineer, who had been a former teacher in the Silverton school and president of the school board from 1883 to 1885. He was, however, defeated by Democrat Dr. James W. Brown, the county coroner "Dr. Jim," brother of "Dr. Bob," Robert H. Brown, a former president of the school board and county superintendent. "Dr. Jim," even the Republican *San Juan* commented, was "a gentleman" whose long experience should fit him for that office.[27] He was soon forced to prove that assertion when storm clouds began to gather over Silverton's school and school board in March 1888.

CHAPTER 7

A SCHOOL IN CRISIS

The recent contest as to whether favoritism and incompetency, or sterling ability and upright intention should obtain in the administration of school matters, has nothing to do with the present order of things. The lady principal is a legally qualified teacher, and all things that can be done to make her administration successful should be done. The opinion prevails in Silverton that only a gentleman as principal can secure the confidence of the pupils and maintain that presence which is necessary to the best results. This may or may not be true, and can only have possibility of proof in the hearty cooperation of the people with the best efforts of the teachers, it being very certain that unless the most favorable circumstances exist, the best results cannot be reasonably expected.

—Superintendent Dr. James. W. Brown in the
San Juan Democrat, September 13, 1888

For Silverton's school the years and months from the spring of 1888 to the fall of 1892 were to bring a series of tumultuous ups and downs in its fortunes. It all began abruptly and totally unexpected in March 1888 when the school's principal, Professor C.M. Kiggins, left Silverton to accept a similar position in Glenwood Springs. When the furor over his departure had died down by the end of the year, two-and-a-half years of calm and happy growth and contentment followed, largely due to the excellent teaching of two sisters, Luella and Mary Burgwin. But Luella Burgwin resigned her position in the summer of 1891, and her sister Mary left town at the beginning of the 1891/1892 school year when the board hired a male principal, John T. Barnett, and demoted Mary from that position. This marked the beginning of another period of upheaval that was to last until the end of 1892. It had been precipitated when the school board fired a beloved teacher in April 1892. The legal repercussions of that event roiled Silverton until the early days of 1895.

The search for Professor Kiggins' successor had started off the school's troubles of 1888. The school board, believing that it represented the town's majority opinion, had looked for a male successor, but was unsuccessful in that endeavor. In the absence of Mrs. Montgomery, their chair, the two male board members present at their March 10 meeting, Secretary W.J. Forsyth and Treasurer Benjamin Austin Taft, then decided to hire another Silvertonian wife, Mercy A. (Mrs. U. Tom) Garrett. Apparently they had been too embarrassed to rehire Kiggins' predecessor, Mrs. Hodges of Silverton. Although they had paid Mr. Kiggins a monthly salary of $133, they now offered Mrs. Garrett $90 per month. They continued to employ Susie Robin as the assistant teacher of the primary grade and upped her salary from $75 to $80 a month.[1] School then opened for the spring term under the principalship of Mrs. Garrett.

As the *San Juan Democrat* reported it, the May school board elections presented "quite a spirited contest" and produced a complete turnover of the board's members. W. Lafayette "Lafe" Henry, the owner of the Silverton hotel, emerged as the new president, C.M. Frazier as secretary, and Thomas Andrew Gifford, a prospector and manager of a Silverton–Red Mountain stage line, as treasurer. At its first meeting on June 14 the new board members, regarding their election as a popular mandate, rescinded the actions of their predecessors and replaced Mrs. Garrett and Miss Robin with Miss Luella Burgwin, a newcomer from Franklin, Pennsylvania. They employed her as principal at a salary of $90 a month. They also voted to reemploy Mrs. Hodges as Miss Burgwin's assistant teacher, "provided she obtain a certificate." But when Mrs. Hodges appeared to be

examined before Superintendent Dr. James W. Brown, she did
not meet his requirements and could not be hired.[2]

The *San Juan Democrat*, its interest aroused by the board's
drastic actions, reported just before school opened on September 3
that Miss Luella Burgwin was a friend and old acquaintance of
Mrs. C.M. Frazier, the school board's secretary's wife. In the
next issue, on September 6, the paper withdrew that statement
"in justice to all parties concerned lest a wrong inference may
be drawn." It did not want to make it appear that the Frazier
family had influenced the hire of Miss Burgwin and the dismissal of Mrs. Garrett. The paper also announced the election
of Frank Prentiss as principal and Luella Burgwin's reassignment
as assistant teacher. Readers were left to assume that the board
members thought these actions necessary to still lingering suspicions of undue influence by the Frazier family. The *San Juan
Democrat* then contributed further to the mystery when it referred to "a slight 'unpleasantness' in public school circles this
week," but assured its readers that Superintendent "Doc. Jim"
is very capably managing these things. He believes that a public
office is a public trust.[3]

Two days after these cryptic reports appeared, the school
board met and, without in its minutes even alluding to
Mr. Prentiss, put Mrs. Garrett back in the principal's place and
charged her with teaching the upper grade for the next eleven
months. Luella Burgwin remained in her place as teacher of the
primary department.

It was then that Superintendent Brown thought it necessary
to enter the scene and show his managing abilities. He
attributed what he called "a partial revolution" in the public
schools to the election of the new school directors. He referred
to Messrs. Frazier and Gifford as "a majority of the self-elected
school board" who had "summarily deposed" Mr. Prentiss and
reemployed Mrs. Garrett. Doing that, they had acted "in accor-
dance with law, as the lady is a legally qualified teacher," but
they had displeased "the sentiment and desires of the people"
who preferred a man as principal. Frazier—whom Dr. Brown
described as "practically the working majority" of the board—
together with the board's president W. "Lafe" Henry—who,
so Dr. Brown commented, "though possibly very useful on

a buck-board, being but a zero annex to the secretary"—
had replied to questioners "with various flimsy excuses and
pretexts," and had defended their choice of Mrs. Garrett as
principal and Luella Burgwin as assistant teacher. The board
members, Dr. Brown thought, still favored their original choice
for Luella Burgwin as principal and Mrs. Hodges as assistant
teacher. But this was besides the point now, and Silvertonians
should acquiesce and give "their moral and practical support to
the efforts of the lady principal."

The *San Juan Democrat* endorsed Dr. Brown's sentiments
and expressed hope that his argument "will have the effect of
cleansing our local educational affairs of political jobbery." A
week later the paper reported that Mr. Prentiss had left town
for Whitewater to direct that town's schools as principal.
Mrs. Garrett could report on November 1 that of the school's
eighty-nine students, sixty-two were enrolled in the primary
grade and twenty-seven in the senior class. When she served out
her term at the beginning of Christmas vacation, the "partial
revolution" appeared to have had played itself out.[4]

At its meeting in December 1888 the "self-elected" school
board members Frazier and Gifford reaffirmed their confidence in Luella Burgwin and, apparently no longer bothered by
suspicions of exerting undue family influence, asked Luella to
telegraph her sister Mary and offer her employment as teacher of
the upper grade for the rest of the school year. The two sisters
then served as the school's faculty at a monthly salary of $90
each. Frazier and Gifford also survived the May 1889 school
board elections, though their president, W. "Lafe" Henry, was
no match for Robert Jay Bruns, a Silverton furniture dealer,
who easily defeated him. The new board quickly acted to extend
school to June 28 with the two Misses Burgwin at their previous salaries. At the board's July meeting the board reemployed
Louella Burgwin for the next school year as teacher for the
primary grade and her sister Mary as teacher of the advanced
grade. It upped their salaries to $95 a month.[5]

For the next two school years the Burgwin sisters presided
in Silverton's public school classrooms with grace and undis-
puted effectiveness. The county's superintendent of public
schools, Dr. James W. Brown, also continued his successful

administration. The *Silverton Standard*, now under the editorship of Charles Snowden, formerly of the *San Juan*, endorsed him in its first issue on November 2, 1889 for another term of county superintendent:

> The public schools have never prospered in Silverton as they have during the past two years, and by re-electing Dr. J.W. Brown, superintendent of schools, the people of this county can rest assured that school affairs will receive the same attention in the future as they have in the past, and our schools will be a credit to the state and country.

Six months later Snowden continued his endorsement of the existing school board as well by urging that Mr. Gifford, whose three-year term as treasurer was coming to an end, be continued in office.

But Mr. Gifford did not choose to run, and his place was taken by Horace Greeley Prosser, the operator of a furniture store and undertaking parlor. He was going to supervise the school together with Mr. Robert J. Bruns as president and the indefatigable C.M. Frazier as secretary. Mr. Snowden was happy. "A better school board," he wrote, "would be hard to find anywhere in the state. All the members have children attending the school and are naturally interested in school affairs and wish to see it continue a credit to the town and to the state."[6]

One reason Silvertonians were happy with their school board was that it had decided to tackle directly the ever-annoying subject of school government and pupil discipline. Their rules, adopted in February of 1889, specified that children afflicted with contagious diseases must not remain in the school; that all children had to be properly clean and dressed; that they walk quietly and make no noise in the school; and that they bring to school books, slate, and other required utensils, unless the board may furnish these as provided in state law. The rules further required written parental excuses for tardiness or absence and set the terms for suspension, promotion and demotion. Pupils were instructed to go home directly from school, not to throw stones or other missiles, and not to mark, scratch, or break books, furniture, walls, windows, or fences. They were held responsible for the repair of any damage they had created.

They were not allowed to smoke or use profane language, fight or expose themselves indecently.

Teachers, too, were covered under the rules. They were held responsible for discipline and should administer it as would

> a kind and judicious parent in his family, always firm and vigilant, but prudent. They shall endeavor on all proper occasions to impress upon the minds of their pupils the principles of morality and virtue, a sacred regard for truth, neatness, sobriety, industry and frugality.

Their teaching would run from 9:00 in the morning to 11:45 at noon and again from 1:15 to 4:00 in the afternoon. There were to be no recesses during these hours, and pupils were expected to leave the school at noon for lunch at home and to return for the afternoon session. The principal was to have power to suspend pupils for gross misconduct or continued subordination and report such incidents to the school board that, by state law, held the power to suspend or expel pupils from school.[7]

With superintendent and school directors having earned the voters' confidence and acclaim, it was the teachers, the sisters Luella and Mary Burgwin, who had become the school's crown jewels. As the 1890s began they were in control of daily operations in the classrooms, and had brought credit to the town by their professional activities. In January of 1890, Luella, who then served as the school's principal, had been to Denver to attend the annual meeting of the State Teacher's Association and thereby had brought a touch of professional recognition to the Silverton school. It was duly noted in the *Silverton Standard*.[8]

The *Silverton Weekly Miner* lauded Luella for her attendance at the Sixth District Normal Institute in Durango in August of 1890. The *Standard,* not to be outdone, praised Mary Burgwin's first-ranked examination results out of a class of sixty to seventy at the teachers' examination in Denver. "This is a greater honor than many people imagine," wrote the paper, "as people congregate from all parts of the state seeking a position in the Denver school, and that a Silverton lady should be in the van is something that every San Juaner may well be proud of."[9]

The papers also reported on the many congratulations the Burgwin sisters had received from parents "on their splendid administration of the school." Board members had been pleased because under the sisters' guidance the number of students who had not missed a single day during June 1890 had increased to fifteen (out of twenty-five) in the upper class and twenty-two in the primary grade. As the school year of 1890/1891 began enrollments had reached 109. Given these encouraging figures and the *Standard's* praise that "today the number of children attending school is larger than it has ever been," the board felt justified in reemploying Luella and Mary Burgwin and raising their salaries to $100 a month.[10]

Silverton's papers continued to portray the Burgwin years as a reign of unalloyed triumphs. The *Standard's* most effusive praise came in its January 1891 evaluation of the sisters' success in increasing the number of children who attended school regularly. The paper recognized that to persuade parents to send their children to school was for nineteenth-century public school teachers and administrators one of their hardest tasks. In Silverton the Burgwin sisters excelled in it. The attendance level "was not caused by force," commented the paper, "but by moral persuasion. It was accomplished by making every child interested in his or her studies, and through the children exciting the interest of the parents." The paper also recognized the readiness of the trustees in furnishing the school with an up-to-date collection of books and maps, and waxed enthusiastic over the children's accomplishments as exhibited in their examination papers. Credit was due, the paper wrote, to the Burgwin sisters. "It was an uphill task for a long time, but now the fruits of their perseverance and energy are showing themselves, and to-day Silverton has a school second to none in the state."[11]

To increase school attendance had been an uphill task despite the passing in 1889 of a compulsory education law by Colorado's Seventh General Assembly. But the law, which applied to all children between the ages of eight and fourteen, had been "inoperative-a dead letter." It compelled the school board "to supply the children of indigent people with clothes," reported Dr. Pascoe, San Juan County's school superintendent in 1893, and that "rendered it null."[12] Seven years later the then State Superintendent of Public Instruction, Helen Loring

Grenfell, complained that matters had not changed much. The law's enforcement, she wrote,

> is made the business of everybody in general and of nobody in particular. It contains a provision relieving from educational responsibility the parents of children living more than two miles from a school house. In the sparsely settled localities of our state it is impracticable to place a school house within two miles of each child; and while most of our people are anxious for the education of their children, and many of them make heroic sacrifices to this end, yet some avail themselves of the weakness of the law and permit their children to grow up in utter ignorance.[13]

In Colorado's mountains compulsory education remained a concept of little everyday relevance.

As May 1, 1891 the date of school board elections approached in Silverton, the appearances of regained stability vanished abruptly. The town's papers reflected a pervasive feeling that while there was much support for and appreciation of the school's teachers, there were many Silvertonians who were out of sympathy with the school directors' involvement in the town's politics. The first indication came with the *Miner's* prediction on April 18, 1891 that Silverton's ubiquitous politician, C.M. Frazier, who had recently been defeated for city attorney, was sure to lose his last public office of secretary of the school board. A week later the paper urged its readers to "arouse themselves" and "let a good man be put up." It reminded them of the role women had played in establishing Silverton's public school and it challenged the town's women to reassert their commitment to strong financial and moral support for the school. "...let the ladies not forget that they have a voice in the matter as to who that man should be." That man, who the *Miner* now termed "the ladies' man," was Mr. Fillmore Hand.[14]

The *Standard* took a slightly different tack. It emphasized that this was "the only election where 'woman, lovely woman' is allowed to vote in this state." And this was true in public education as in all political fields. As the state superintendent had explained in his *Seventh Biennial Report*, only qualified electors could fill county offices, and they were confined to men. Thus, he wrote, "a woman cannot fill the office of County Superintendent of Schools."[15] But it was different on the school district level. So the *Standard* then did not just join the *Miner* in supporting "the ladies' man," but proposed that "to make it interesting it would be well to nominate some lady for the position."[16] The lady in question was to be Emma Hollingsworth, who in the winter of 1878/1879 had been the town's teacher and throughout many years the leader of the town's literary discussions circles, chronicler of community events, and author of many obituaries.[17]

Then, a week later on May 2, two days before the election, the parents of Silverton's schoolchildren were shocked by the announcement in both papers of Luella Burgwin's resignation as teacher of the primary department. She had accepted, effective the last day of April, the position as principal and primary teacher of the public school at Red Mountain. The *Miner* commented that "our people have naught but kind words, and the best wishes of the town follow her to her new field of labor." The *Standard* noted that "every one will be sorry to see her leave." There was no indication in the papers what had prompted her departure after three years of devoted and supremely successful service.

The attention of the papers' editors, however, was focused on the coming election, and the papers now took diametrically opposed positions. The *Miner* saw itself as champion of what it called a "long-sought and badly needed reform in school matters," a reform that would take school issues out of political intrigues, place less emphasis on economy, and show greater concern for uncrowded classrooms and academic achievement. As more reform candidates had appeared and threatened to splinter the reform cause, the paper dropped its support for the ladies' cause and Mr. Hand, the ladies' man, and backed a petition, signed by over 100 citizens, to unify the reformers behind Mr. Alonzo Smith. Smith, the *Miner* wrote, is "a man of culture and intelligence," who had for many years been a school director in the northern part of the state.[18]

The *Standard*, too, fell silent on the concerns of Silverton's women and gave up its support for Emma Hollingsworth, because, the paper explained, someone had said: " 'Whoa

Emma' and she withdrew." Instead the paper came out for a local businessman, Fred Helmboldt, a candidate who it thought would represent conservative and economical school management. The *Standard* liked him because "he has floated no petition upon a long suffering public and does not want the position. The office is seeking him..."[19] When the votes were counted on May 4 the *Miner*, the reformers and their lady supporters had to admit defeat. Mr. Helmboldt was the new board secretary and was to serve together with Robert J. Bruns as president and Thomas Andrew Gifford as treasurer. Economy had won out over reform.

The new board wasted no time in hiring Miss Mary Ross as replacement for Luella Burgwin in the primary department. Mary Ross had taught before in Farmington and Durango, and was introduced to the community at school closing exercises for the primary class. Both papers commended the progress and good behavior of Miss Ross' students and featured Mary Burgwin for turning the exhibition exercises of her upper class into a successful fund raising drive for buying an encyclopedia for the school. At its next meeting the Board began to debate the hiring of teachers for the next school year. In its deliberations about a principal it turned again to a man as their candidate and invited Professor John T. Barnett of Potsdam, New York, a graduate of the New York State Normal and Training School. It thereby demoted Mary Burgwin from that position and assigned her, an experienced upper grade teacher, to the primary department.[20]

The reaction was not long in coming. Mary Burgwin, the *Miner* reported, resigned from Silverton's schools and left for California. A second heavy blow had fallen on Silverton's parents and pupils. Both Burgwin sisters were now gone. "The prize they think they hold so firmly, ... the talented and beloved principal of the Silverton school," the *Standard* lamented, had slipped through the grasp of Silverton's people. "It began to dawn upon the parents and children most nearly interested that, in vulgar parlance, they were left."[21]

But worse events were yet to come and threw the school again into turmoil. At its meeting on April 1, 1892 the school board decided that with the increasing number of school children in town they should add another teacher and classroom. They turned again to Mrs. Garrett, who had served as principal in the spring and fall terms of 1888, and asked her to teach the primary department. For the intermediate room they assigned Miss Ross who, they said, should in the absence of Mr. Barnett also act as principal. The *Standard* greeted the decisions effusively and told its readers that they should "rejoice that they can now offer their children the best advantages for a good common school education." Now their children would receive greater personal attention and parents were given the opportunity "to keep the children who are now running the streets and absorbing all the deviltry that they can, and they can absorb a heap of it, in school and in their proper place."[22]

A week later, the paper reversed itself 180 degrees. It had come to its and the town's notice that on the preceding Monday, April 4, school board members Bruns and Helmboldt with Principal Barnett as their executive officer had come into Miss Ross' classroom and ordered her and her advanced students to leave and continue teaching in another room. Miss Ross complied but expressed her unhappiness over this abrupt change. Principal Barnett and the two directors then asked her to resign which Miss Ross refused to do. Thereupon the men unceremoniously dismissed Miss Ross from her classroom. The reason, the paper surmised, was that having received negative reactions from many Silvertonians for having hired two additional teachers in addition to Principal Barnett, board members were concerned that president Robert J. Bruns might be turned out of office at the next election.

The *Standard* itself did its best to further that cause. The hiring of Mrs. Garrett and Miss Ross, it wrote, "is evidence that they [the school directors] intended rather to bring themselves and the school into ridicule than anything else." It called one member "a barnacle on every society or board it has ever been his misfortune to be associated with," and the other "a poorly tempered tool in the hands of a 'scab' workman." These men had fired Miss Ross "without a shadow of a cause, or even an attempted excuse..." Thus, the *Standard* concluded, "one of the most satisfactory teachers ever employed in our schools was peremptorily and unlawfully ousted." There was only one lesson to be drawn from this: School board president Bruns must be replaced by "a man...who will look to the interests of the school, instead of his private spites and who will know something of what can lawfully be done."

The town reacted to the news with an outpouring of support for Miss Ross. Miss Ross published a note of appreciation and thanks to the parents of her pupils in the *Standard* and announced that she would continue her teaching in a private hall for all her pupils who would wish to come. She would not charge any tuition fees. The *Standard* on its part assured its readers that it bore no grudge against Principal Barnett whom it considered "a friend of ours" and who had acted, not on his own, but as an employee of the board. "It is the school board we object to," the paper repeated, and it was Mr. Bruns' defeat by J. Frank Molique in the unusually packed board election of May 2 that it celebrated.[23]

Under its new chair the Board then rehired Mr. Barnett as principal and Misses Luella M. Liscomb and Sarah H. Dyson as teachers for the next school year. Having in mind the public's distaste for high teachers' salaries, board members paid Mr. Barnett $100 per month and the two teachers $75 per month each. When school began on September 6, Miss Dyson had been replaced by Miss Louise Walling, a graduate of the New York State Normal School. Miss Liscomb, who had graduated from the high school in Cameron, Missouri, was to teach the primary class, and Miss Walling, in charge of the intermediate grade, was also to teach vocal music in all three grades. The board also had decided to furnish free textbooks for all students and thus, commented the *Standard*, there was no more reason for parents to keep their children out of school.[24]

Had it not been for what came to be known as the Ross case, the time of turmoil for San Juan County School District No. 1 might now have come to an end. But the board was faced with the unpleasant prospect of having to deal with a $300 suit filed against it by Mary Ross. Board members resolved to defend themselves even though they knew community sentiment was on Mary Ross' side and many had urged that they let the matter rest. Claiming that Miss Ross had taught without a teaching certificate, the board filed a counter suit for $800, the sum it had paid Miss Ross as salary. The court, however, rejected the board's claim and the jury awarded Miss Ross $300.[25] Silvertonians were delighted. The story spread through the region and was gleefully told by the Santa Fe *New Mexican*. The *Silverton Standard* could not resist reprinting it:

Over at Silverton, Colorado, a Miss Ross—who by the way, judging from the noise the people of her town have made over her, seems to be very pretty and popular and a good deal of a peach—had been fired out by a heartless School Board three months prior to the expiration of her contract. Then Miss Ross took her case to the courts. Five days were required to try it, and when the jury brought in a verdict awarded Miss Ross $300 and 8 per cent interest the town went wild. The local brass band was called out and serenaded the young woman. She made a speech and the people cheered and built a bonfire. That's the way they do things in the West for fair womankind.[26]

The board, however, remained obdurate. It received and placed on file a taxpayers' petition to pay the court's judgment and laid on the table a motion to appeal the case. A week later they decided to go ahead and appeal the district court's verdict to the state supreme court. Thereafter in neither the board's minutes nor the newspapers do we find any further mention of the Ross case until, two years later, the school board minutes of December 26, 1894 reveal that the board paid $ 215.00 for the judgment and $38.70 in interest. When two month later in February 1895 the board paid an additional $53.50 in court fees the case finally was laid to rest.[27] The years of turmoil in the school had ended.

What lay behind the trials and tribulations that beset Silverton's school community during the late eighties and early nineties? Superintendent Dr. James Brown tells us that we should look to the preference of many Silvertonians for a male principal. The ever recurring reports in Silverton's papers over pranks, mischief, and "deviltry," as they called it, committed by youngsters on the streets support that view. An authoritative male voice was needed to call youngsters to order and persuade parents to send their youngsters to school. That view led the board in 1891 to hire Professor Barnett and lose the much beloved Mary Burgwin.

But there were those in Silverton who heartily disliked taxes and demanded economy in the financing of the school. A male principal, a professional with professorial title, was more expensive than a woman principal who, in several instances, was a fellow Silvertonian and was paid only slightly more than a teacher. This, I suspect, led Messrs. Forsyth and Taft in 1888 to hire Mrs. Mercy A. Garrett as principal at a monthly saving of $43.

Again others, and especially parents of school-age children, supported what came to be known as the reform party whose positions were usually broadcast in the *Miner*. The reformers were willing to hire more teachers, supply additional teaching materials, and in general oppose their economy-minded fellow citizens. They suffered most when the Burgwin sisters left Silverton, and they supplied the cheering section for Mary Ross. But economy-minded voters, whose views appeared in the *Standard*, also supported Mary Ross and her case when they were loath to have the board finance further appeals.

However one may see these issues, one thing is clear: the low points of Silverton's school turmoil can in every case be attributed to the sometimes real and sometimes presumed political and personal interests and intrigues of Silverton's school board members. The high tides of Silverton's school fortunes flowed from the devoted and successful work of the school's teachers, the Burgwin sisters furnishing the outstanding example during this period. No incident illustrates better the contrasting approaches to their respective tasks than the dismissal of the conscientious and effective Mary Ross by school directors whose deliberations were swayed by financial rather than educational considerations.

CHAPTER 8

THE TURBULENT 1890S

The outlook for mining in our county is good. Now is the time to bring before the "gold bugs" of the east, the great gold producers of San Juan County.... There need be no great alarm. The fall of the price of silver is a terrible calamity to the American people, especially to the silver producing states of the west, but the condition can not continue very long. Our people have too much good sense and will show it in a way that will change these conditions.

—The *Silverton Standard*, July 8, 1893

The ups and downs of Silverton's public school during the late 1880s and early 1890s were played out against the background of turbulence in the town's economic and political affairs. National and worldwide developments affected life and mining enterprises in the San Juans as they had never before. There was no way in which Silverton could have escaped them. Yet throughout the decade its newspapers, while neither ignoring nor denying the effects of the nationwide depression of 1893, kept up an optimistic outlook: There was gold and silver in "them thare [*sic*] mountains," and those resources would insure the town's economic survival.

At the beginning of the decade the Superintendent of the Census had declared that "at present the [nation's] unsettled area has been so broken into by isolated bodies of settlement that there can be hardly said to be a frontier line." But there never had been a line that divided settled and unsettled land in the Rocky Mountains, and mining camps and towns such as Silverton had sprung up wherever gold and silver ores had drawn eager prospectors, businessmen, and financiers to erect their cabins and bring in their mills and smelters. The closing of the frontier, understood by many as a warning of or threat to America's ongoing economic growth, had little meaning to the people of the San Juans.

Yet economic news from across the country sounded dire. There were more strikes reported in 1890 than had occurred in any previous year. By 1892 the Homestead strike at the Carnegie Steel Company near Pittsburgh provoked bloodshed between workers and Pinkerton detectives and made Westerners wary and afraid of labor unrest among its own miners. To be sure, in 1878 Congress had passed the Bland-Allison Act that had directed the Treasury to buy between $2,000,000 to $4,000,000 worth of silver every month and had led to the coinage of 378,166,000 silver dollars. The act, however, had proved to be ineffective. The Congress stepped in again and in 1890 passed the Sherman Silver Purchase Act that required the Treasury to buy 4,500,000 ounces of silver monthly.[1] Though these measures appeared to have paid heed to Western demands for the free and unlimited coinage of silver, President Cleveland's announcement early in 1891 that he opposed the free silver cause,[2] provoked disappointment and resentment among Silvertonians over the national Democratic party's drift toward support for eastern industrial interests. When, then, in the summer of 1893 President Cleveland attributed the financial panic of 1893, the collapse of many railroads and banks and the resulting nationwide unemployment to the effects of the Sherman Silver Purchase Act and Congress repealed the act in November of that year, a sense of gloom settled over the West.

The next year Coxey's Army, a band of unemployed industrial workers, set out from Ohio to march on the Capitol in Washington to deliver their petitions of protest. Arriving, they were promptly arrested for trespassing. In the farming areas of the Midwest, the Populists, gathered in the People's Party, gained strength rapidly, so much so that they drew Silver

Democrats to their ranks and brought about the split of the Democratic Party.

The climax came in the national party conventions of 1896. The Republicans, meeting in St. Louis on June 16, rallied around the gold standard, and nominated William McKinley of Ohio. Their apparent unanimity was broken only by Colorado senator Henry M. Teller who, declaring his departure from the party, led twenty-two silver delegates out of the hall. In the Democratic convention at Chicago a month later, the Silver Democrats dominated the scene. Nominating William Jennings Bryan, the "Boy Orator of the Platte," they captured their party, and caused the Populists in their convention to also declare Bryan as their standard bearer. The country now had a clear choice: free silver and a holding fast to a populist agrarian order or the gold standard and the acceptance of industrial and financial capitalism. In the election of 1896 the voters decided for the latter.

As Silverton entered the 1890s its economic fortunes appeared promising. "The mining district," the *Silverton Weekly Miner* wrote, "is looking so well this season that we frequently hear that Silverton has seen her worst days, and that a gradual and continued improvement in all branches of business is confidently expected."[3] The town installed water in the county jail and school house and by the end of September the Silverton Electric Light and Power Company had turned on electric lights on the city streets and in the houses that had been wired to receive the service.

One reason for the upswing in business had been the passage of the Bland-Allison Act and the Sherman Silver Purchase Act. In Silverton the *Miner* rejoiced: "Men with money are coming in and are prepared to develop properties which for a decade have been neglected...and this year will yet witness some fine new discoveries, as well as the uncovering of large ore bodies by the working of now idle properties." As 1890 came to its end, the *Miner* listed the accomplishments. Freight rates had been reduced, prices for lead and silver had risen, new mills had been built on North Star and Silver Lake and others, and output had increased in the mines on Red Mountain and the Animas District.[4]

The good times appeared to be continuing throughout the next two years. In February 1892 the *Standard* boasted of new ore discoveries all over the county. "The San Juan is holding its own," the paper wrote, "and when all the railroad and real estate excitements have vanished, this county will still be increasing its shipments from year to year, and forging ahead to the position of the greatest producer in the state." When then in May 1892 the *Standard* reported the reopening of the Bank of Silverton and at the beginning of 1893 the legislature was petitioned to build a new wagon road from Silverton to Durango via Castle Rock Springs there seemed reason to be proud and confident.[5]

That confidence, however, was not an unmixed blessing. It provoked neighboring towns to belittle Silverton's achievements, and Silvertonians resented that. As the *Standard* told it, reporters in Durango to the south and in Ouray to the north made stories "out of whole cloth" that "the superabundance of snow" had led to scarcities of such staples as flour, sugar, coffee and fresh meat. "It is not pleasant," complained the paper, "for people in Denver and the east to read that their friends and relatives are snow bound and starving, when such is not the case." To the contrary, the paper assured its readers, the staples mentioned and others such as "ham, bacon and canned goods can be had in any quantity."[6]

And prosperity had its shadowy sides as well. Gambling and liquor sales flourished, wrote the *Miner*, and "ladies are daily insulted on our streets by drunken men as they puke forth their ribald jests." The *Standard* joined in and reported "about fourteen fights in town Wednesday night." A year later the paper reported the case of a "Chinaman" convicted of opium smoking by the testimony of eight or ten of his colleagues. It wondered about "the white people who claim to be more enlightened than the Chinese but who supported this 'joint,' and asked: 'Are they less guilty than the Chinaman?'"[7]

Prosperity also had boosted the fortunes of the "Sage Hen" Jane Bowen. Besides running her dance hall, saloon, and bordello she and her husband William also owned mining properties and ranked among the wealthiest families in town. When William died of miners' consumption in 1891 Jane sold all her properties and, as the *Standard* reported, Silverton's "old land mark" planned to leave for her old home in London. But by the summer of 1892 she was back in Silverton and with a grand ball opened a new saloon and dance hall, the Palace Hall on Twelfth and Blair Street.[8] She could not resist the lure of Silverton's prosperity.

San Juan County's mining entrepreneurs, however, were not entirely satisfied with the federal government's silver legislation. They demanded nothing less than the free and unlimited coinage of silver. Mine and bank owner Edward G. Stoiber and the ever-present C.M. Frazier agitated for the creation of a Free Silver Club in Silverton and shaped the club's resolutions.

> We are opposed, [they wrote,] to the present tyrannical dictation of the money lenders and capitalists to the congress of the United States to debase silver, the money of the laborer and money-borrower, and we demand of that August body its complete restoration.

Frazier ridiculed those who wanted to use the country's silver for "tea-sets, sugar bowls and backs for looking glasses." Proudly calling himself a Silver-Republican, he said, he stood with the wage earners and condemned "the grasping, full grown, kite shaped gold bug and capitalist."[9]

A year later a different tone appeared in the paper's writings. The *Standard*, still claiming the outlook for San Juan mining was good, urged the county's gold producers to remind the "'gold bugs' of the east" that silver was the lifeblood of the West and that "our people" demanded its free coinage. That editorial was followed a week later by a mass meeting in the city, demanding the free coinage of silver and condemning any adoption of a single gold standard. The Congress, however, repealed the Sherman Silver Purchase Act, and the price of silver tumbled from 80 to 60 cents per ounce.[10]

As the reality of the depression of 1893 set in, the *Standard* sought to whitewash the situation as much as it could. In spite of the crisis people still came to Silverton to enjoy the scenery and cool mountain air. Those who left would come back as soon as the price of silver would recover. The town trustees would restrict employment to men with families only, "and let the single men rustle somewhere else." There would be "dull, hard times" when everyone must cut expenses and those who oppressed their neighbors "should be summarily dealt with." Then, giving vent to dammed-up feelings, the paper wrote that to argue with eastern journalists "who put the production of silver on a par with the production of potatoes," and were unteachable about monetary questions, "was like trying to educate an asylum of idiots." The only good that the *Standard* could see as advantage of the hard times was that now "we can all wear out our old clothes."[11]

What brought home the seriousness of the depression was the growing unemployment among workers across the country. The Silverton Silver League spoke up in defense of Coxey's Army. The League condemned Republican senator Edward Wolcott for having denounced the marchers as "a menace to our institutions" and for having claimed "that in Colorado or anywhere there is work for all who want it..." Both Silverton's weekly papers spoke out in defense of the unemployed and hungry. In early July of 1894 about 150 miners gathered in Silverton at the Town Hall to discuss the formation of a miners' union. Again the Silverton papers expressed their support: "The right of labor to organize may be questioned by some of the grasping corporations but by honest people never," wrote the *Standard*.[12]

The *Miner*, usually an outspoken Democratic and labor-friendly paper, looked for any evidence it could find to soften the dire threats and consequences of the depression. It found it in what were the growing discoveries of valuable gold deposits in the San Juans. In early 1894 the paper wrote that "when the great depression attendant on the decline in the price of silver struck with terrible effect in the mining camps of Colorado, San Juan County scarcely felt the shock at all...[I]t was discovered that many of our largest mines, hitherto producing nothing but silver, were rich in gold...It is believed," the paper wrote, "that should silver become worthless that her [San Juan's] gold mines could better sustain her population than her silver mines have done in the past....What town can compare with Silverton?," it asked. Otto Mears was building a third railroad for Silverton and, the *Miner* rhapsodized: "There is peace, comfort, plenty and prosperity in San Juan County."[13]

The Silverton papers but reflected the general trend of business affairs in the San Juans during the mid-1890s. As Duane Smith writes in his *Song of the Hammer and Drill*, "mines shut down or curtailed production, laying off miners and adversely affecting Colorado's entire economy…, wages were reduced,…business…declined sharply…followed by inevitable failures and bankruptcies." But the papers' encouraging prediction that gold mining was to carry on the San Juans' mining operations where silver might falter, turned out to be true. Between 1890 and 1899 gold production shot up in value from $1,120,000 to $4,325,000 while silver remained at slightly above $5,000,000 throughout the period. By 1899, wrote Smith, "silver was more a by-product of gold mining than strictly its own master." In the San Juans gold had softened the impact of the depression.[14]

Nonetheless, Silverton's town board continued to wrestle with unemployed miners who, the *Standard* wrote in May of 1895, "are now in town and a menace to our public welfare." By November a strike threatened to break out at Edward G. Stoiber's Silver Lake mine over the company's announcement that it would impose a $1 per month dues for services to injured miners at the Durango hospital. But within a week the strike threat evaporated when the Durango Sisters of Mercy offered "to supply competent medical care, the necessary medicines, board and nursing to all who are injured or sick, and in all cases where it is necessary, to have the services of a physician." The miners agreed to pay the $1 fee to assure transportation of the injured to the hospital in Durango.[15]

By year's end a tone of regained confidence prevailed. The Del Norte *Prospector* summed it up when it wrote that the coming mining season of 1896

> promises to take the lead and will probably precede a genuine stampede to San Juan….In the face of the hardest sort of times for silver mines, San Juan has forged steadily to the front and will continue to do so. Harder times have evolved closer and more business like methods in mining, and who shall say that a direct benefit has not accrued to San Juan as a result?[16]

Silverton's politics reflected the up- and downswings of the county's economy and widespread unhappiness over suspected corruption among Republican office holders. In the city election in April of 1890 Democrat John Wingate had been reelected as mayor against the opposition of what the *Miner* called "the saloon element, and its tin horn following; the Frazier group...and...the non productive and non-taxpaying people." As the state elections approached in the fall divisions in the electorate appeared that crossed the usual party lines. The *Silverton Standard*, normally a Republican leaning paper, voiced its distaste for what it called "the straight Republican ticket." Taxpayers should vote for it, the paper told its readers, if they wished "to see the taxation thermometer rise," and parents should support it if they wanted to see "the school lands stolen from their children's children." The *Standard* opposed the reelection of Fred Dick as state superintendent of public instruction. It accused him of having engaged in shady manipulations with the school land fund and asked whether the people of Colorado proposed "to vote men into office for the purpose of stealing the heritage of their babies?" It was alright for Republicans to vote for Nathan B. Coy, the Democratic candidate. The *Standard* also spoke out against C.M. Frazier who this time ran as candidate for state representative. "The decent element of the Republican party are disgusted with the ticket," declared the paper, "and well they may be."[17]

The *Miner* happily agreed and then blasted Frazier as one who, as school board member, was suspected to have exacted commissions from school salaries and who, "as an official of the people...had in every instance worked against the interests of the taxpayers and played into the hands of the corporations." He did not stand for republican principles and "honest Republicans have been browbeat long enough and are prepared to snow the chronic under." Women, the *Miner* predicted, would not vote for him because they did not like his record on the school board." The *Standard* corroborated that judgment when it reported that a vote had been taken among schoolchildren that gave Frazier's opponent, Democratic candidate Sullivan, a large majority. "There must have been considerable interest in politics among the ladies," concluded the *Standard*.[18] When the results came in they showed that Mr. Sullivan had won the house seat and Democrat Nathan B. Coy was the new state superintendent of education.

In the town elections of the following spring, the *Standard* greeted the victory of the People's party with its mayoral candidate George L. Thorp as a decisive defeat of Mayor Wingate's economy policies which, so wrote the paper, had left sideways and crossings sadly neglected. The *Miner* was delighted to report that C.M. Frazier had lost his office as city attorney and now served only as secretary of the school board. He would lose that office, the *Miner* correctly predicted, in the coming May 10 school board election. The women's vote would defeat him.

But the *Miner* had underestimated the indefatigable C.M. Frazier's vote-getting capabilities among men. As president of the newly created Silverton Gun Club Frazier got his name in the *Standard* at each of the club's first three August meetings. He emerged twice as the best shot and, at the third meeting, he managed second place. When he let it be known among his friends that he was interested in the position of county attorney, the Commissioners handed him the job in February 1892.[19]

The depression and the women's issue then proved to be dominating topics in the state elections of 1893. The *Standard* again blasted the straight Republican ticket with Frazier at the top. "What can you expect of a party that will tolerate and follow after the lead of a rattle-brained monstrosity like Frazier? Will the honest and respectable voters allow themselves to be dictated to by a lawyer, whose sole object is notoriety, and a whack at the boodle crib?" It championed the People's Party and its candidates, among them Dr. J.N. Pascoe, who was nominated as the candidate for county superintendent of public instruction.

The main issues, however, were women's suffrage and free silver. The *Standard* saw them as intimately related. "A vote for equal suffrage is a vote for free silver. The man who goes to the polls on the 7th of next November and votes to exclude half the ballots of half the citizens of Colorado from the ballot box on account of sex, is a traitor to the white metal and to the best interests of the State." We can be sure that our "culture bearing" and school promoting Silverton ladies had done all they could to persuade their husbands to vote for the suffrage ticket. As it was, the Republican ticket never even appeared on the Silverton ballot, and the People's Party only local rival was the Silver Ticket 16:1. The People's Party won hands down, with equal suffrage being approved by a margin of ninety-six votes. Statewide, women suffrage had prevailed in most counties in which the People's Party had carried the day. Counties with Republican majorities, the *Kit Carson County Banner* wrote, had "assassinated" the women's issue. The Populists, the paper added, had proclaimed "to open mouthed Republicans that their mothers, wives, sisters and daughters were, at least, equal to the Hottentot."[20]

Political fortunes are fickle, and the city spring elections of 1894 brought the Citizen's Party with its Republican candidates back into power. That happened despite the *Miner's* exhortation to its readers: "Down with the rotten hearted, money hoarding, sack loving miser-whelps, terming themselves free coinage republicans," and despite the *Standard's* reminder to its "lady voters" that they "should remember that it was the People's Party that gave them the power to vote in this state." The *Standard* was furious. Not only had its favored candidates lost, but black crepe had been hung on their door knobs and whoever had done it, "was not a fit person for any civilized community."[21]

Perhaps the Republicans won because their most controversial would-be candidate, Cassius Marcus Frazier, for once did not run for office. He and his wife had closed down their business and all their properties and, as the *Miner* told its readers, "folded tent and like the Arab silently stole away." The *Miner* testified that Mr. Frazier had been a successful lawyer whose political ambitions, however, had never been gratified. They had left him with "blasted, withered hopes. Socially," the paper wrote, "Mr. Frazier was a pleasant companion, a conversationalist well versed on the topics of the day.... [H]ad he possessed that personal magnetism so essential in a public man, success would have crowned his efforts." In Silverton, it was not to be, though in subsequent years Frazier continued his political ambitions and was elected attorney-general of the state of Arizona.[22]

The Republicans' victory in Silverton was not to last. A year later, in the spring of 1895, the People's Party emerged as the winner in the city contest with John Casey being elected Silverton's mayor. In the fall election the Populists put up a woman as candidate for Superintendent of Schools, Mrs. Ellen Carbis. She easily defeated the Citizen's ticket candidate, Dr. J.N. Pascoe.[23] Not only was Silverton again safely in Populist hands, it also had demonstrated the power of the newly won women's vote and had put into office the first woman superintendent of public schools in San Juan County. Ellen Carbis was to be reelected three more times and held this office until January 1904.

The second half of the 1890s with the discovery of gold had brightened the outlook for Silverton's economy. The *Silverton Weekly Miner* summed it up at the beginning of 1897. The town's population stood at 2,000 and during eight months of the year San Juan County's mines employed over 1,300 men, two-thirds of them during the winter. The town proudly showed off an electric light plant, a waterworks system, two banks, three churches, and fourteen lodges. There were no paupers, the paper claimed, and there had been only two business failures in the past ten years. Of special importance for the town's economic fortunes was the completion of several new railroads. Otto Mears' Silverton Railroad to Red Mountain and Ironton had been completed in 1889, his Silverton Northern to Howardsville and Eureka in 1894, and the Gold King Mining Company's Silverton, Gladstone and Northerly Railroad to Gladstone in 1896. The town "can now boast of being one of the great railroad centers of southwestern Colorado." With the Durango-Silverton connection, it "has now four railroads..." triumphed the *Miner* a few weeks before the new century opened.[24]

Silvertonians, determined to leave behind bad memories of the depression, made the most of their regained prosperity. During the winter months, home entertainments, dances, and church socials crowded the calendars of Silverton's adult society and helped participants to forget the anxieties of the past. Once the snow was gone, bicycling became the newest fad that took the town by storm. Apparently it was not a hobby of the town's youngsters only and did not count as one of the "deviltries" boys had been accused of. To the contrary, "old men and young men, small men and large men, ladies of sedate and dignified mien, all have the craze," reported the *Miner*. People practiced on the town's baseball diamond that at times took on "the appearance of a training course for amateur circus riders." The event that had everyone "oh" and "ah" was Professor Harris' balloon ascension that took place in the evening of July 21, 1898. The professor floated in "the aerial monster" up to 400 feet and then descended to where he had started from. Hardly had he climbed out of the gondola when "the huge mass of inflated canvas" took off on its own, stayed in the air for about half an hour until it crash-landed in a back alley. Silvertonians would remember the event for many a day.[25]

Nationwide developments in corporate control of industries and labor strife both brightened and clouded the picture as many of the smaller mines were sold to big companies and absentee mine ownership became more common. As investors expected to profit from the application of new techniques, such as the use of electricity, the telephone, power drills, compressors, the safety cage and improved tramways from mine to mill, Silverton benefited from eastern and foreign capital flowing into the San Juans. Writing of the 1890s, Duane Smith summarized it: "Consolidation, corporation growth, and the switch from silver to gold distinguished this decade."[26]

But there was a downside as well. Newspaper reports reached Silverton in September 1896 of martial law and bloody battles with strikers in nearby Leadville, and during the summer of 1897 of great coal mine strikes in the Midwest.[27] As labor and industry became embroiled in these battles, the *Silverton Standard* accused the nation's judicial system of unconstitutional usurpation of a citizen's natural and legal rights. "Now our judges declare that labor has no rights that capital is bound to respect," wrote the paper. "This is a departure from the fundamental principles of Republican government and until late had no precedent in American history." The *Miner*, speaking up for the Democrats, chided Republican senator Wolcott for having "forgotten about the strikes and lockouts of the past summer when speaking of our present prosperity..."[28]

Locally, the papers brought up familiar issues such as complaints over dirty alleys, unsafe street crossings, dimly lit streets, and a missing sewer system. In winter they kept the town informed of avalanches and untimely snowy deaths. A story with a happy ending was that of mail carrier John Bell who had been reported killed in an avalanche on the Red Mountain–Ouray route near Ironton. He had lain unconscious for twelve hours in an ice cave and, wrote the *Standard*, when he woke up and extricated himself he "lived to read his own obituary." The *Standard* also showed its less appealing side of journalistic racism when, reporting that a colored barber and porter was given forty-eight hours to leave town, it added that it was "needless to state that he left." At another time it excoriated unemployed and vagrant miners for patronizing "heathen Chinese restaurants" because there they would get "more grub" for their money. It told them: "You look to white people for a job and ask a living price for your labor and then go and spend your money with the scum of God's creation."[29]

The soiled doves of Blair Street, too, made their regular appearance in the papers. When "Ollie" had died, a repentant girl living, as most of the bordello residents did, under an assumed name, two ladies had laid a cross of flowers beside her. The *Miner* commented piously: "Perhaps the screwing down of the coffin lid over the poor unfortunate's face is the happiest fate that could befall her." At a later time when "Emma," another Blair Street resident, had died from alcohol abuse the attendants wanted to know her real name to inform her parents. But "Emma" refused, preferring, as the *Miner* put it, "to die among strangers without a friend to shed a tear over her bier." Not all Blair Street departures were final. "Mabel" had swallowed diluted carbolic acid, but with the help of a stomach pump and a gag had been brought back to consciousness. "Jealousy is said to have been the cause of her action," reported the *Miner*.[30]

Local issues were overshadowed in 1899 with the statewide battles over the eight-hour law. The Colorado legislature had approved it on March 1, 1899, for work in mines, smelting and ore reduction works, and it was to become law on July 1. Mine and smelter owners fiercely opposed it, and moves were underfoot to get the Colorado Supreme Court to repeal it. The owners of the Durango Smelter, involved in that attempt, coordinated their efforts with several of the larger Silverton mines. As a result these mines closed in June, throwing 600 men out of work. The *Standard* called it "a temporary staggerer right in the face of the bright and cheerful outlook of increasing prosperity on which San Juan County has based its aspirations for many a month." A month later the smelter owners announced that after July 10 they would again purchase ore and run their smelter "whether or not the Supreme Court holds the 8-hour law constitutional."[31]

The Court on July 22 did then in fact declare the law unconstitutional, but in the meanwhile mines in Telluride and Ouray and other Colorado locations had continued or resumed work on the eight-hour schedule. In Silverton, however, an uneasy standoff prevailed. Contradictory rumors circulated that the Durango smelter was about to open up or that it owners were contemplating to sell it off; that it was running full blast, but that the miners' union had lost the strike. The *Miner* reported that some Silverton merchants refused to extend credit to the out-of-work miners and that the Durango smelter, under guard

Figure 8.1   Ross Smelter, Silverton, Colorado

by eighteen deputy sheriffs, was trying to get under way with a workforce of about thirty men.[32]

By late August the difficulties were resolved. The Durango smelter was back at work and the San Juan Mining Association had agreed that underground work would resume under the eight-hour system. "Let us all rejoice," wrote the *Standard*. The *Miner* closed the subject with a reaffirmation of its advocacy of organized labor "through thick and thin," and of its optimistic prophecy of Silverton's future:

> Never before since the inception of Silverton as a mining center has the outlook appeared so bright....Our labor difficulties having been laid away on the shelf never to be recalled, our principal mines and reduction works being operated under full blast, new mines being developed, new mills rapidly approaching completion and unlimited capital drifting in here for investment, therefore we can but reiterate the assertion made above.[33]

There was no longer any reason to have doubts about the future.

CHAPTER 9

SILVERTON'S LIFE AT CENTURY'S END

At the beginning of the present school year a State Scientific High School was established in Silverton. The course of study was enlarged co-equal with the preparation of the State University and the pupils of the high school course are admitted into the State University or State Normal School without examination. The course of study is almost a verbatim copy of the State course. The high school is a strict embodiment of the State University Preparatory Latin Scientific Course. Two additional teachers have been employed and a third will soon be necessary.

—*The Silverton Standard*, January 2, 1897

The most important event for Silverton's school and people during the closing years of the century was the introduction of a high school curriculum. For Silverton's parents this meant that if they planned for their children to attend a college or university, they no longer had to send their boys or girls away from home to a private academy or public high school. For a town to establish its own public high school was to bring to a triumphant conclusion its struggle for the completion of public education. In Silverton the agitation to introduce a high school curriculum had begun when in February 1895 the school board had finally managed to close the books on the Ross case and its aftermath. Now the slowly but steadily growing number of children of school age in city and county demanded the school directors' attention. At the beginning of the 1894/1895 school year enrollments stood at about 150 students. The grammar department instructed by Principal W.D. Cunningham consisted of thirty-one students in grades five to eight. The intermediate group taught by Miss Mae Jones counted thirty-six students in grades three and four. The primary department under the instruction of Mrs. Mary B. Hodges held fifty-eight students in grades one and two. Six years later the picture had changed. A high school had been added, the teaching staff consisted of six teachers, including principal, assistant principal, and high school assistant, and the number of students had risen to 220.

These changes had done little to allay the ongoing disquietude among residents over their school board and the various principals the board had installed. They had in no way lessened the reciprocal exasperation felt by school board members and principals over the way they were often viewed by the public and portrayed in the press. At the time of the 1894 board elections the *Miner* had reminded its readers that it was now more than ever important to elect competent and fair-minded directors who stood "aloof from all petty striving," were "unswerved by party clique or influence," and could "estimate at its true value the advice of vicious meddlers, who always know how to run other people's business better than they do their own." The *Standard* shared similar sentiments. It reported from New York that the mayor of Brooklyn had appointed five women to its board of education who, so a New York paper had written, "will at least tend toward cooling the atmosphere of the board room on meeting days." The *Standard* suggested that similar results might be obtained in Silverton.[1]

Principal Cunningham, the teachers, and some parents were unhappy about lacking citizen and board support for a much-needed addition of a classroom and a teacher to better serve the growing number of students. Cunningham was convinced that Silverton ought to add high school instruction to the existing three room graded school which consisted of grades one and two of the primary department, grades three and four of the intermediate department, and grades five to eight of the grammar department. "There is no reason on earth," he wrote in the *Miner,* "why we cannot have as good results here in our schools as in any high school in the state and graduate a class every year well equipped to enter the freshman class in any college." But a public that accepted "scores of boys playing on the streets in plain view of every one," he complained, lacked the enthusiasm to support the work of its teachers.[2]

At the end of the school year Cunningham vented his anger and disappointment in his annual report. He had to deal with the ever-recurring citizen complaints about lack of discipline in the school and of youngsters playing on the street during school time. But when he tried to enforce discipline, he knew of parents who threatened him for punishing pupils. He thought the harshness of Silverton's winter and spring was responsible for widespread absenteeism, but he had to admit that thirty-five students had dropped out during the school year. His teaching methods were blamed for that. He defended them, saying that he was but following "the ideas of our most prominent modern educators and training schools." Like many another progressive educator he complained that he had been met with parents' and the public's lack of faith and displeasure and with his students' disorder and outright refusal. He tried to convince Silvertonians that the object of the public school was not general subject information but the stimulation of a symmetrical development and unfolding of the children's faculties and powers. "Character is the great object in the training of the citizen," Cunningham wrote, not the "mere process of cramming knowledge into the brain..." He closed his report with a plea for Silvertonians to establish a high school of their own. He had already initiated a ninth grade with four students who would present their finishing orations at commencement exercises. "Let this be only an indicator of what can be done if the proper means are furnished." Only one more teacher was needed, and Silverton's youngsters would have "a firm foundation for any occupation" or could "enter most any university in the country."[3] Cunningham here referred to the then debated views whether public high schools should prepare their students for occupational life through modern subjects or whether they should prepare them for college entrance.[4]

Principal Cunningham must have known that Silvertonians did not appreciate his lecture. Three days after his report had appeared in the paper, the school board elected a new principal, Professor F.M. Vancil of Jacksonville, Illinois, and voted that

> if the principal or anyone of the teachers should fail to govern or control the pupils during school hours in their several departments in a proper and orderly manner, the Board shall deem it cause for removal, a majority of the School Board being the Judges thereof.

To add insult to injury, the *Miner* reported that at the graduation ceremony of Mr. Cunningham's four ninth-grade young ladies, Miss Mae Jones and Mrs. Hodges were highly praised as their teachers. No mention was made of Mr. Cunningham. Later, in August, the *Miner* carried a brief notice that Professor Cunningham had been chosen as principal of the high school at Mancos. "We wish him success," was the *Miner's* brief farewell.[5]

With Mr. Vancil as principal, Mrs. Hodges as teacher of the primary department, and Sadie L. Knowlton of Pennsylvania taking Mae Jones' place in the intermediate department, the new school year began on September 3, 1895 with 140 pupils showing up. The board pleaded for citizen support for the school now venturing into high school work. There would be no lack of discipline, it promised. Regular and prompt attendance would be insisted upon, and students would not be allowed to study just what they wanted. The *Miner* approved of these remarks, but kept a watchful eye on school developments. When the board decided the day before school started to hire a fourth teacher, Mrs. Jennie M. Reavley, for the grammar department, the *Miner* immediately informed its readers of the costs involved. Now school expenses would rise to $406 per month. Noting the new teacher's Irish name, the *Miner* wondered how she had secured her position when there was a strong American Protective Association (APA)[6] lodge in town. The *Standard* countered with approval and support and less suspicion. It opened its pages to the monthly school reports and to "School Room Echoes," occasional contributions by the students reporting on school events. It noted with approval the arrival of a thirty-four-volume encyclopedia which the school board had purchased for the school library.[7]

The school year of 1896/1897 then brought momentous changes. A high school formally began its work. The State University preparatory department's Latin Scientific course was adopted as its curriculum. Its graduates would be admitted without examination to the State University or State Normal School. Silverton's public school now had four departments: The principal's department, including the high school, taught by Mr. Vancil and his assistant, Miss Mabel Daniels; the grammar department under the direction of Mrs. J.M. Reavley; the intermediate branch, instructed by Mrs. Mattie B. Carpenter; and Mrs. Mary B. Hodges' primary department. As usual, enrollments fluctuated throughout the year. They ranged from a high of 216 to a low of 177. Silvertonians appeared to be content with the school's progress, and the *Standard* congratulated the school on the high quality of its female teaching staff. It was critical, however, of teachers enforcing discipline by encouraging pupils to tell on each other. Such teachers, the paper wrote, are not suited for their work.[8]

But there was opposition also to the high school, chiefly because of the increased costs. The *Miner* saw to that. As the May 1897 school board elections approached a spirited contest broke out between the incumbent secretary John Rogers and his challenger, Charles E. Robin, the owner of an insurance, real estate, and abstract business. The *Miner* took Robin's side, writing that he had been persuaded to run "by the urgent wish of a multitude of the citizens." The *Standard*, speaking up for John Rogers, wondered why people who had shown little interest in education now sought to secure a major vote "even unto the corralling of spoiled doves off Blair Street for the purpose of electing a manager for the most sacred institution of all—our public school?" A few days later, when Mr. Robin had won the election, the *Standard* noted that not only had the school board candidates elicited a good deal of comment, but also "a good deal of sharp talk" was going on about the school's teachers. There was "a constant kick about this and that of minor importance much of which is generally laid to the fault of the principal," who, the paper reported, had accepted a new position elsewhere.[9]

The *Miner* soon enlightened the town what the "sharp talk" was all about. The newly elected school board member, Charles E. Robin, also known as "Bob" Robin, opposed the hiring of married women as teachers. The school could do better with one less teacher than it employed now, he let it be known, and discipline was not well enforced. "A little more tickling with a strap and less with the fingers would be beneficial," he reportedly had said. The real culprit, however, the *Miner* announced, was the departing principal. It called him "one of those spectacles of humanity which make the very atmosphere depressing" and "a sordid and contemptuous character, worse than a miser who would starve his stomach to fatten his purse." How could the old school board have hired him for a second year? "A man who received over $6 a day, lived on twenty cents, and wore a straw hat the year round, is unfit for a high school promoter and the undertaking here was an absolute fizzle, despite the columns (self written) in the Standard." Principal Vancil had not even bothered to attend the school closing exercises and had instead taken the train to leave town. "The school board did the correct thing in deducting this time from his bill for services."[10]

So the board was in need again of a new principal and found him in Mr. F.E. Mullen. While all the school's teachers for the 1897/1898 year were women, Mr. Robin was pleased to notice that, in contrast to the preceding year, none of them were married. The board also had listened to the demand for lesser school costs and had reduced the principal's salary to $110per month and all of the Silverton teachers to $80 per month. Mrs. Hodges refused to accept her reduction in pay and was replaced with Miss Cora Edwards as the teacher of the primary department. The board agreed to open an ungraded school in Howardsville and to pay its teacher, Alice Hendrickson, a monthly salary of $35. On the ever-recurring question of truancy and tardiness, the board decided to allow each teacher to use her own methods.[11]

As the board and the school got ready for the school year 1898/1899 the *Standard* thought it strange that the board secretary, Charles E. Robin, known for cutting teachers' pay, now was voted by his board colleagues an annual salary of $50. It also thought it a "great wrong" that when the board declined to rehire Principal Mullen for the next school year and chose Professor Edward L. Howett of the State Normal School instead, it invited "from a distance, men to fill a position for a term of one season, especially so when one with a family in that period only becomes fairly settled and acquainted with the people." Once the new principal arrived he persuaded the board that if he were to teach a tenth grade with only four students enrolled he would have to neglect those in the lower grades. The board then decided to consolidate the advanced students in the ninth grade and to eliminate the tenth grade for the time being.[12]

As yet another school year began in the fall of 1899 the board was again faced with the dilemma of having to find a new principal. Mr. Howett had accepted the position of principal in the Animas City school. The board replaced him with Mr. Elmer E. Amsden, a graduate of the Iowa State Normal School and of the Colorado Normal School at Greeley. Primary teacher Cora Edwards also left Silverton for a teaching position in Billings, Montana, and Miss Norah Greamba took her place. All teachers were required to sign a contract that committed them to treat all children alike and in no event to show partiality to any child. The board also decided to build a new school house at Howardsville and to reestablish the high school's tenth grade. It agreed to compensate principal Amsden and assistant principal Virginia McMechen for any extra time that they might spend in their additional duties.[13]

As in the past, the board continued to admonish parents to show greater interest in the school. Only three parents out of 200 had shown up at the Arbor Day exercises. "I should think," wrote board member Louis Wyman, "you could realize how disgusting it is for a child to study and learn a recitation and then have to speak it to an empty house." Visit the school, the board asked, "do not wait for a special occasion. Enter the rooms without knocking." Write excuses for children's absences from school only when they were sick or the absence was unavoidable. The board warned Silverton's tobacco dealers not to sell tobacco in any form to children under sixteen. It caused great annoyance to teachers, and criminal penalties would be pursued. The *Standard* complained about the "devilment" committed by boys. They had recently cut wiring and extinguished lights in town. It also reported that the school board was considering erecting a new school building to make room for the high school grades ten and eleven. And, no doubt having in mind past comments of the *Miner,* the *Standard* implored Silvertonians to keep politics and personal prejudice out of school board elections.[14] The papers asked Silvertonians to enter the new century with a new and more supporting attitude toward their school. After all, theirs was one of the few high schools on Colorado's western slope. Their children could now finish their schooling at home. Should they not be proud of it, cherish, protect and strengthen it?

As for the children themselves, Robert Redwood's monthly report card of 1897 allows us a glimpse into the academic schedule of the high school that had begun to operate during the preceding year. It shows us that Robert was a well-behaved student and got high marks in grammar, physiology, and physical geography, though he did not make the grade in Latin, algebra, arithmetic, and U.S. history. Lisette Fast Robinson reminisced of her years as a teenager in Silverton during the 1880s. The town, she wrote, was "a little center of civilization imported from the east." She remembered the railroad that "traversed the narrow and picturesque canyon to Durango" and the school. She, too, had taken Latin like Robert Redwood, and "four years of English literature and mathematics, though it didn't progress beyond algebra." She remembered her male principal who, she thought, had been attracted to the frontier where "things happened." Her high school training, she wrote, "was as efficient and remarkable as any" that she had heard of taking place in later times.[15]

While Silverton's newspapers kept the town's citizens stewing over the high school issue, they also kept alive the ever-recurring controversies over the presence of and goings on in gambling houses, dance halls, and bordellos. The city council had sought

SILVERTON HIGH SCHOOL.
MONTHLY REPORT; Principal's Department.

Report of *Robert Lockwood*

TERM COMMENCING AUGUST 30, 1897.

STUDIES, ETC.	1	2	3	4	5	6	7	8	9	AV.
Days Present.	17½	17½	0	11	19	13				
Days Absent.	1½			1	1	6				
Times Tardy.	0			0	0	1	0			
Deportment.	88			78	80	70	70			
Arithmetic.					70	73	65			
Algebra.	70									
Geometry.										
Grammar.	88				6-	78	74			
Rhetoric.										
Civil Government										
Physiology.					96	89	78			
U. S. History.					70	70	60			
General History.	78									
Geography. Physical					88	90	79			
Zoology.	83									
Botany.										
Geology.										
Eng. Literature.										
Amer. Literature.	78									
Physics.										
Latin.	72									

This card will indicate the standing of the pupil in each study pursued, regularity of attendance and deportment. All rank is on a scale of 100. Less than 75 is unsatisfactory. Parents are requested to examine and sign this report each month.

T. E. Mullen, ...Principal.

Figure 9.1   Silverton High School Report Card

once again to diminish immoral influences by prohibiting women to dispense or purchase spirituous liquors and intoxicants. But the *Miner* pointed out that as long as liquor licenses restricted sales to no special sex and as long as women were enfranchised and eligible to hold office, so long city authorities could not deprive them of their rights and privileges of American citizens. That was true, the paper stated, even when the intent was

> to lessen the profligate inclinations of unfortunate depravity....So long as the city fathers license the selling of intoxicants within the abodes of lewdness they cannot justly restrict to these subjects and objects the right of purchasing where they please without working an injustice upon certain individuals connected with the liquor trade.

And what had been the result of bearing down on gambling in the city the paper asked. "The city has lost the better element of gambling and the money involved and it still retains the most obnoxious element in creation—the worthless and unmanly incapable of self-support."[16]

The *Miner* also denounced the anti-immigrant and anti-catholic principles of the American Protective Association, a nationwide organization that flourished for a while during the 1890s. Using a quote from the *Durango Democrat* that, in turn, had copied from the *Commercial Appeal* of Memphis, Tennessee, the *Miner* suggested that the letters APA stood for American Protective Asses. The *Miner* called it "un-American, un-constitutional, and un-Christian," an organization that proclaimed "that the pope is about to chaw up the constitution and run off with the government...[and that] the country is going hellward at a mile a minute because Catholics are allowed to vote and hold office!" The *Miner* later added its own reflection that "the A.P.A. will doubtless still run its course of local mischief for some years. The embers of religious prejudice, bigotry and intolerance are never quite extinguished..."

The paper likewise spoke out for free speech and academic freedom. It defended the professors of Brown University in Rhode Island who had protested the firing of their President Andrews by the trustees. It quoted Senator Teller who, as told in chapter 8, had left the Republican Party at its 1896 convention

in St. Louis and had now declared the trustees' action as
"Un-American." Teller had vowed that he would never return
to his party as long as it refused "to denounce any attempt at
the strangulation of free thought, the debasement of colleges,
and the destruction of liberty as I see it going on."[17] For the
Miner, Teller's pro-silver and pro–free speech views made him
the favorite spokesman for Colorado liberals.

For national and local politics, the year 1896 had brought
the momentous confrontation between the agrarian and silver
standard crusaders, inspired by William Jennings Bryan, and
the business and corporate interests who had made their political home in President William McKinley's Republican party.
Bryan had rallied his forces at the Democratic convention in
Chicago with his battle cry against the proponents of the gold
standard: "You shall not press down upon the brow of labor
this crown of thorns, you shall not crucify mankind upon a
cross of gold."[18] These ringing words had not failed to give
renewed impetus to Colorado's silver forces. With the news
of the Leadville strike and its bloodshed still on their minds,
Coloradoans had gone to the polls and had given their votes
to the Democrats and William Jennings Bryan. In Silverton
Bryan headed the ballots of five other parties: the National
Silver Party, the National People's Party, the People's Party—
also called the Populists—, the Silver Populist Party, and the
Silver Republican Party. The Democratic victory swept into
office Alva Adams as governor and another woman, Miss Grace
Espy Patton, as state superintendent of public instruction. The
country, however, had voted for William McKinley and the
Republicans.[19]

Once the excitement of the national elections subsided,
Silverton politics returned to the familiar local debates between the Citizens' and the Peoples' parties. "What does it
matter to us in this city," asked the *Miner*, "whether...candi-
dates for aldermen, mayor, etc. are Democrats, Republicans, or
Populists?" The paper opened its columns to the Citizens' Party
and saw it and its mayoral candidate Charles H.H. Kramer win
the city spring elections. The *Standard*, having seen its candidate, F.O. Sherwood, go down to defeat, found fault with what
it called the "personal ambitions" of the new administration

and predicted that they would prevent the construction of a new city hall. It promised to continue supporting the Populists who were "conscientious in their desire to continue the routing of gangsters."[20]

The *Standard* then could take great satisfaction with the outcome of the county elections in the fall of 1897. The *Miner*, forgetting or discarding its earlier dismissal of national party candidates, had complained that "Silver Republicans, Populists, Silver Democrats and Silver Populists are Democrats who do not know their names," and that they ought to vote the straight Democratic ticket. To no avail. All candidates on that ticket lost, and the Populists and Silver Republicans reelected Ellen Carbis to the county superintendency of public instruction.[21]

In the spring city election of 1898, the Populists continued their victories. The Democratic mayoral incumbent, C.H.H. Kramer was defeated by the People's Party John Casey. The chief issue had been the Democrats' effort to support a bond issue for the city to own and manage its own public light company. "The vital question in this election," the *Miner* had announced, "is that of the bonds–whether Silverton will longer submit to the extortion of the electric light company, their poor service, their exorbitant rates and their insufficient light..." Silvertonians were not impressed with that argument. They were not willing to saddle the city with bonded indebtedness and defeated the bond issue eighty-three to forty-three. The People's Party now ruled in both county and city.[22]

In the state's fall election, the Populists were part of the allied silver forces together with the Democrats and the Teller Silver Republicans. They had nominated Charles S. Thomas for governor and Helen L. Grenfell for superintendent of public instruction. In Silverton Thomas Annear was the candidate for state senator on the *Standard* supported People's Party ticket. He was opposed by the Democratic candidate John T. Barnett, the former principal of Silverton's school. The *Standard* urged his readers to vote the straight Populist ticket and choose Annear because he had promised not to assist Senator Wolcott or any other "goldbug" for U.S. senator. When the votes were counted, the Populists had again emerged as victors in Silverton.[23]

The Populists' string of victories would continue for one more year. Supported again by the *Standard* in the mayoral spring election of 1899 the Populists, allied with the Democrats, Republicans and the Citizens' Party, put Frank Brown into the mayor's seat. The *Miner* had returned to its advocacy of local parties and now spoke up for the newly created Independent Party. Again the public ownership of the light company, and now also of the water works, was on the ballot. This time the *Standard* urged voters to endorse the bond issue. "Let us own both electric light plant and water works, not alone for the purpose of lessening the cost to the consumer, but to make more adequate the supply," it wrote. Yet Silverton's voters remained unconvinced. Though they had favored the *Standard*-supported united parties, they shied away from authorizing bonds. All the *Standard* now could say was that the city pay the light company well while protecting businesses and residences from extortion. The *Miner* expressed its hope that, if its franchise were renewed, the water company would improve its plant. The light situation would remain as it had been.[24]

In the fall election of state officials the Populist winning stream came to a halt. When the *Standard*, professing that it preferred candidates selected for their business rather than their party affiliations, promoted a fusion ticket of Populists and Republicans, many Populists became suspicious. The *Miner*, pointing out the strange coalition of Populists committed to Bryan and Republicans endorsing McKinley, thought that these Populist voters "realized to what ignoble purposes the People's party was put in San Juan county, and the revolt was natural." Democratic candidates won the election, except for Ellen Carbis who was reelected as county superintendent for public instruction on the Populist ballot.[25]

As soon as the new year had begun, a debate resumed over the public ownership of water works and electric light plants. The *Standard* remained noncommittal. "The people are going to demand...a system of water works that will furnish water whether it be owned by city or corporation," it wrote. It is not a question of public ownership but one "of having at all times a supply of water..." The *Miner* saw it differently. Under its new ownership of Henry L. Damschroder it supported the issue of bonds and the Democratic candidate for mayor, former school director Charles E. Robin. Again the Democrats won, Robin defeating his opponent, Josiah Watson of the Citizens' Party, by 372 to 250 votes. The water bonds this time carried the day on a 110 to 50 count. The *Standard* admitted defeat: "Now comes the question of municipal ownership of light and sewers. Might as well study up the question and fall in line, for it's bound to happen."[26]

During these years Silverton's political battles had also been shaped by events in the Carribean and Pacific and the nation-wide debates over imperialism, war, and peace. A revolution against Spanish rule had broken out in Cuba in 1895, and on February 15, 1898 the American battleship *Maine* had been sunk by an explosion in Havana harbor with the loss of 260 American sailors, one of them, Doc Lambert, a native Silvertonian.[27] A wave of outrage swept over the country, and *Remember the Maine* dominated headlines and talk everywhere. By June American troops took off to liberate Cuba, authorized by a congressional Joint Resolution with an added amendment by Colorado senator Teller that declared the United States had no intention to exercise sovereignty over the island. A Spanish fleet was sunk in July near Santiago Bay, but the greatest victory occurred earlier in the Pacific when in late April Admiral Dewey destroyed a Spanish fleet in Manila Bay. By the end of July, the United States dictated its peace terms to Spain: relinquish Cuba, cede Puerto Rico, and accept American occupation of city, harbor, and bay of Manila.[28]

In Silverton the *Standard* enthusiastically waved the flag and beat the drum for the war. It excoriated President McKinley for his lack of patriotism and sense of duty. It faulted him for having indicated in his war message of April 11 that Madrid had capitulated and was willing to negotiate. Though the president was willing to intervene, the *Standard* wrote, he did not explain "whether it is to be done through the Wall Street stock market, or by prayer and an invoice of Sunday school tracts..." A few days later it asked whether Silverton's school was in sympathy with Spain because it did not fly an American flag. On the same day it reported that when McKinley had signed the Joint

Resolution and sent Spain the ultimatum to evacuate Cuba by Saturday noon

> the enthusiasm of our populace was intense. War talk waged hot and fierce and an impromptu parade was gotten up, headed by the Silverton Cornet band. Flags waved from all the business houses and flag staffs. 'Twas a gala day for the right, and that day will always be memorable in the hearts of those who are crying for vengeance in behalf of the Maine and her gallant crew and for liberty on the part of the 'Spartan band' struggling for life and freedom against their Spanish oppressors.

If there existed "war fever" in Silverton, the *Standard* did its best to keep it at a high pitch.[29]

Much as the *Standard* then supported the war measures and praised President McKinley for his "masterly way" of having conducted the campaigns and much as it approved of the "gold and also silver Republican press" glorifying "Major McKinley for the military and naval tactics by which he put down Spain," the paper had reservations. It could not forget that in his war message McKinley had suggested that the United States accept Madrid's capitulation to the American demands. It therefore warned against elevating the president "to any position to which he is not justly entitled." He had "merely done his duty which has been compulsory." A real war president would not have shown his indecision and lack of fighting spirit.[30]

The *Standard* added to its disapproval of President McKinley a note of sarcasm about the country's business interests whose pro-imperialistic propaganda was driven by the profit motive. It cited with approval the *Nebraska Democrat's* derisive statement:

> We are going to annex the Philippines and Christianize the natives. The good work has already commenced. Milwaukee has already shipped 489,000 bottles of beer to those benighted heathen. As soon as their heads grow happy and their hearts become soft under the benign influence of the beer we can send them some missionaries and bibles.

A year later it printed William Lloyd Garrison's *Onward Christian Soldiers* with its concluding lines, "How natural that a change should come in nineteen hundred years, And Bibles take a place behind the bullets and the beers!"[31]

But it continued to belabor McKinley for his "long drawn out dilly-dally policy in the orient" that, by refusing to send to the Philippines "a proper force of soldiers" had only prolonged the conflict. Still, the *Standard* maintained, the Philippine campaign will restore peace to the islands. It "will not mean their enslavement but as far as possible their enlightenment..." The Filipino people will adjust themselves "to new conditions in a way that is upright, wholesome, proper and American in the highest and the truest sense."[32]

During this entire period, the *Miner* confined itself in the main to factual reporting of the events as they unfolded, though it also left no doubt that it opposed the ongoing war in the Philippines. It spoke out "against slaughtering the best blood of American youth for the purpose of extending the field for Standard Oil and other trusts." It linked business support for the war to the attempts of Silverton businessmen to aid the smelter trust by refusing credit to striking workers. For the *Miner*, business interests and the "gold bugs" were responsible for war and economic exploitation at home and abroad.[33]

CHAPTER 10

SILVERTON ENTERS THE
TWENTIETH CENTURY

In these times of strikes and dissensions in mines, smelters, factories and other business enterprises, it is no little satisfaction to be able to point out a prosperous mining community with all its mines and mills working to full capacity and with no thought or fear of strikes or labor troubles. Silverton, in southwestern Colorado, has been one of the most peaceable and contented mining camps in that great mining state, and it gives every promise of continuing along its even way without trouble or disturbance. There may be several reasons why this happy condition prevails there, but the principal reason, undoubtedly, is that in that particular camp both mine owners and miners have been reasonable in their demands, and have been willing to live and let live.

—Quoted from the *Chicago Chronicle* and printed
in the *Silverton Weekly Miner*, July 8, 1904

As the new century's first decade got under way Silvertonians looked at mining, their town's mainstay for economic survival, with mixed emotions of fear and smugness. From all around them in the San Juans and Colorado mountains came news of labor unrest, strikes, lockouts, and dissension. Troops had been called out to protect mines and mills, and vigilantes had brought gunfire and bloodshed. But Silverton itself remained quiet. Its mines continued producing, their yearly output steadily rising in value (table 10.1). Where once silver had reigned supreme, now gold led production, followed by copper and lead, and silver ranking fourth.[1]

The business aspects of mining, however, were changing. Corporations were taking over once individually owned mines. Managers directed local operations for owners who lived in eastern cities or foreign countries. The Guggenheims of New York bought Edward Stoiber's Silver Lake mine with its boarding house, mill, power plant, and tramways. Inevitably, Silverton's papers portrayed the issue in the light of party politics. The voters, wrote the *Silverton Standard*, will have to choose between Democrats, who champion "the hopes of the masses," and Republicans who represent "the brutality of the classes." While older established mines such as Red Mountain's Yankee Girl declined and ceased operations others such as The Gold King, Sunnyside, and Silver Lake prospered. The net effect was positive. By 1909, so Duane Smith tells us, "the San Juans claimed a place as one of Colorado's top mining regions."[2]

Despite their own good fortune of relative labor peace Silvertonians could not but be disturbed by the news of strikes and violent clashes between miners and national guard troops. Nearby Telluride had been the scene of strikes in 1901, the assassination of a mine manager in November 1902, the use of gunfire and dynamite, the appearance of national guard troops in November and December of 1903 and again from March to June 1904. Other strikes had occurred at Cripple Creek and elsewhere. The *Silverton Weekly Miner* called the Cripple Creek event a sympathy strike "and therefore unjustifiable." It sided

Table 10.1   Value of Ore Produced in Silverton Mines

Year	Value of Ore (in Dollars)
1899	2,100,000
1900	2,200,000
1901	2,934,904
1902	3,048,170
1903	3,245,607
1904	3,320,170

Source: From Annual Reports in the *Silverton Standard*.

with the mine owners. "The wrongs, or alleged wrongs," the *Miner* wrote, "from which the miners in Colorado suffer have been committed in other parts of the state, and the mine owners in the Cripple Creek field, except in a few instances, have had no hand in them." The *Standard* reported in September 1903 that 1,000 troops had been dispatched to Cripple Creek, that a strike had broken out at the Durango smelter, and that a walkout had occurred at the mills in Telluride. It condemned "mob rule" in Telluride where "200 good 'law abiding' citizens [had] arrested sixty-five miners and shipped them off to Ridgway from where they walked to Ouray. Some of them might have been agitators," observed the *Standard*, "but they were quiet and orderly. No reason for mob rule," the paper concluded. When the unrest finally subsided it appeared that the end result of the confrontations was the defeat of the Western Federation of Miners at the hand of corporate and management power.[3]

Except for a brief strike at the Gold King mine Silverton's miners had continued their employment. Most of the district's miners belonged to one or another of the local unions that had retained the support of the community. They had been able to lend aid and assistance to their striking brethren in other cities.

> Silverton takes a hand in helping out the striking miners, [wrote the *Standard*,] in order that they may be able to hold their own against the coal baron, the gigantic railroad trust and those who say that the miners will work for the starvation price offered or suffer the consequence of a close down of the monopolized coal mines of Pennsylvania.

The town was "too substantial a camp to be permanently crippled by any strained relations on the labor situation."[4]

The *Miner* was likewise satisfied with Silverton's labor conditions. They were

> the very best in the state as employee and employer are well satisfied and are working under signed agreements between the Miner's Union and the San Juan Mine Owners' Association which agreements are faithfully kept by all parties working thereunder [*sic*]. Very few idle men are found in the county ...

Nobody wanted a strike.

> If there is any person in this camp who wants the miners to
> strike, they are not good citizens, and are working against the
> best interests of the community. If any work is being done
> towards causing a strike, it is done in the dark, where cowards
> always work.[5]

For Silvertonians the most ringing endorsement of their
productive labor conditions came from the *Chicago Chronicle*
when in July of 1904 it described the town as "one of the most
peaceable and contented mining camps...with no thought or
fear of strikes or labor troubles." Silvertonians could be well
satisfied with themselves.[6]

There had been other encouraging news aside from the
absence of labor troubles. During 1900 the Kendrick-Gelder
smelter had been built at the mouth of Cement Creek and in
December it started to operate. In September, the *Standard* had
reported, "two hundred men are on the pay-roll at the Stoiber
terminal and new mill site, a large force is at work on the water
plant and before many days a still larger force will begin work on
the new sewer system...." In the next month a Silverton-Denver

Figure 10.1   Silverton, Colorado, 1901

telephone connection had been established. New faces appeared on the city's streets. Some were welcome, the paper remarked, others not. Among the former belonged tourists who, so the *Standard*, enjoyed the sunshine of the Rockies "while their friends are sweltering throughout the east." The latter included hobos who looked for grub and a place to sleep and shunned labor. They, wrote the paper, "should be made to 'hike' down the track."[7]

The *Miner*, proudly boosting its union credentials, reported that Silverton's printers had begun to organize a union and the paper would welcome their efforts. It promised "to grant any reasonable concession the union may demand," but surmised, that "the [R]epublican sheet down the corner will probably have a roar coming." Even outsiders confirmed the town's rosy prospects: "Silverton is enjoying a greater prosperity than it has experienced in years," wrote the *San Miguel Examiner* in August, "and the mining industry is very active. Operations this year are on a good systematic basis and the properties will be greatly benefited thereby. Considerable Boston capital is invested in the camp." The next year the *Miner* happily noted that "more than fifty new houses have been erected in Silverton this summer, many of them handsome, substantial brick and stone buildings that would be a credit to any community." What the town still needed, the paper wrote, was a new hotel and a courthouse.[8]

But not everybody was as happy. The *Standard* wrote a year later that

> there seems to be an opinion prevalent on the outside that Silverton is booming. Nothing could be further from the truth. It is, like it has always been, a thriving camp surrounded by resources very attractive to the capitalist and prospector, but there are plenty of men in the camp to fill industrial wants and the worker who comes here looking for a job is taking big chances of not finding it. There are plenty of idle men in Silverton and will be until the floating population attracted here by reports of a boom quits the camp.[9]

Those who were skeptical of Silverton's prospects found further confirmation of their views in the renewed failure of the Bank of Silverton and the suicide of its president, James H. Robin, on new year's day of 1903. It turned out that Robin, known in Silverton as a successful businessman and public spirited citizen—he had been a brick yard, sawmill, and lumber yard owner, a prospector and operator of an abstract business, as well as a banker and had served as Silverton's town treasurer and trustee and as deputy county clerk—had diverted bank funds into the portfolios of mining companies of which he was the major stockholder. His body was found near the tracks of the D.& R.G. railroad.[10]

One subject that would again and again agitate Silverton's people was the controversy over public versus private ownership of utilities. In the spring of 1900 a debate had begun over the municipal ownership of a city water system and both city papers, the *Standard* and the *Miner*, supported the issue. The *Standard* warned Silvertonians that the Silverton Water Supply Company was not in a position to deal with a major fire in town. "The people are going to demand . . . a system of water works that will furnish water whether it be owned by city or corporation." The question that then appeared on the April ballot of the municipal elections was whether the city should buy or build a municipal waterworks plant and system and issue bonds for the purpose. When the votes were counted, the bond issue had passed by 110 to 50 votes.[11]

In the September 1901 election the question then was whether the city's water system was to be owned by the city. Again both papers endorsed the proposal, and the voters agreed to municipal ownership by a vote of forty-eight to nineteen. Referring to the nineteen opposing votes, the *Standard* commented that there were a few tax-paying citizens "who are so extreme in their non-democratic views that they think everything, even the government itself, must be controlled by corporations."[12]

Next came sewers and electric lights, and the two papers took different positions. Although both praised the idea of a municipal sewer system, the *Miner* held out for a private franchise for an electric light company. That was heresy for the *Standard*, and it condemned it as a graft that had "no feature commending it." The *Standard* felt gratified when the voters in the April 1901 turned down the franchise by a vote of 113 to 74.[13]

In the March election of 1902 the question of a municipally owned electric light plant came up for debate. Again the *Miner* remained reluctant to join the *Standard* in an unqualified endorsement. It thought the town to be too heavily indebted to extend municipal ownership at this time. Sound business consideration, the *Miner* wrote, persuaded it to forego the Democrats and to support the Citizens' ticket. That coalition, the *Miner* wrote, was nonpartisan and "municipal government is business, not politics." Yet the *Miner* had to concede defeat when Silvertonians voted 106 to 64 for municipal ownership of the electric light plant. By the end of the year a municipal light plant was under construction, and one year later the *Standard* praised the benefits of competition that had persuaded the privately owned Silverton Electric Light and Power Company to lower its rates and to "keep the city plant within the present bounds of reasonable charge..." As far as the *Standard* was concerned, everything from the city's water and sewerage systems to its light plant was first-rate. "Silverton is an ultra-municipal ownership town."[14]

The city's commitment to public ownership was reflected also in its voting records in support of Democratic candidates. In the national elections of 1900 Silverton went solidly for Bryan. As it had been in 1896, the silver issue was foremost on everyone's minds. "There is not an individual in San Juan County but realizes what free coinage of silver will be to this section," wrote the *Miner*. "The lines are squarely drawn....Take your choice; are you a Gold Bug or a Bi-metallist?" The *Standard* did not disagree with its competitor. "Bryan's speech of acceptance has the true ring of Democratic principles," it wrote. "His words for humanity and against imperialism; his appeal to right the wrongs done and being perpetrated by the money powers, his praise for all liberty loving people...place him far above the politician and tricksters of the type of [Republican] Hanna and his lobby."[15]

When the votes were counted Colorado had voted for Bryan and for James B. Orman, the Democratic candidate for governor. In San Juan County Bryan had beat McKinley by

1,128 to 362 votes. Yet McKinley had won nationwide. "We mourn," lamented the *Standard*. "Bryan is lost beyond recovery under the wave of McKinleyism.... Colorado stood nobly to the front to down President McKinley, yet a state with but four electoral votes had but a mighty small say so in the matter.... We bow to the inevitable."[16]

Democrats also dominated the electoral scene in local contests. N.A. Ballou enjoyed three consecutive terms as mayor that stretched from the spring election of 1902 to the spring of 1905. When in 1903 he won over the Socialist candidate Caspar Malchus who was being supported by the *Miner*, the *Standard* had asked its readers:

> When it comes to a question whether Democracy, the cradle of all American political reforms, or Socialism shall rule in Silverton, which is already conceded to be the most advanced in municipal government of any small city in the United States, there ought to be no hesitancy among fair-minded voters. Shall demonstrated practical ability govern the city for the ensuing year, or shall Silverton be turned over to the dreamers?

The results made it clear that practical ability had won over the dreamers.[17]

Ballou's luck, however, ran out in the 1905 contest when in an otherwise Democratic victory he lost by two votes to his Republican adversary Fred Goble. This time it was the *Miner's* part to enjoy its triumph: "The town election on Tuesday developed some surprises, especially for the other fellow. The idea of a Republican being elected mayor had never entered their heads and they are still dizzy when they contemplate the results." The *Standard* nonetheless could still take satisfaction from the fact that all other contested seats had gone to the Democrats.[18]

Election results for other officials were similar during this period. In 1902 San Juan County voted for the Democrat E.C. Stimson for governor but Republican James H. Peabody garnered the majority of the state's vote. The next year the county voted for Democrats for Supreme Court judge, justice and constables. In the 1904 battle for president between Theodore Roosevelt and Alton B. Parker San Juan County voters again chose the democratic ticket, as they did in the governor's race between Republican James H. Peabody and Democrat Alva Adams. Although the country went for Republican Theodore Roosevelt, Colorado elected Adams as its governor. But election irregularities forced Adams to give up his seat and persuaded Peabody to resign. The governorship then was taken over by the Republican Jesse F. McDonald. "A new era has opened up," rejoiced the *Miner*. "McDonald is governor without an enemy in either party. Now let us have peace within and without the party."[19] For the *Miner* the state and national triumph of Republicanism was ample reason to express delight.

Silverton's thriving economy and its political struggles played out against the background of a lively, not to say tumultuous, social life in which prostitutes and their opponents as well as ethnic minority folks provided the major and not always voluntary actors. The "good people" of Silverton never had been very sympathetic to the Blair Street sisters, though they knew how to esteem them for the contributions they made to the town's public purse. The *Standard* started off the new century advising the city authorities to consider a choice between "eternal vigilance with no increased night watch force" or "perhaps a burglary or two." After all, Silverton was "rapidly acquiring a very tough looking lot of individuals, who, like the lilies of the valley, 'toil not, neither do they spin.' "[20]

To make matters worse, as the new year began the town was threatened with invasion and who knows what kind of destruction by the formidable, hatchet wielding, six foot tall and 180 pounds weighing Carrie Nation who billed herself as "a bulldog running along at the feet of Jesus, barking at what he doesn't like." Silverton's saloon keepers were alarmed. They had made preparations to let down stout barricades in front of their windows should a signal be telephoned from the depot that the temperance crusader and her cohorts had arrived. "Our police department will not interfere in any sense of the word," the *Standard* assured its readers. Apparently the mere threat of Mrs. Nation's arrival had its effect. The city council spoke out "in regard to women frequenting saloons" to curb "such open and flagrant violations of ordinary decency" as the "inmates of Blair Street" were accused of. Whether or not as a result of the council's action, there were no further reports of Mrs. Nation's threatened or actual arrival.[21]

But reports of the doings of the Blair Street sisters continued to appear in the papers. On September 7, 1901 citizens could read in the *Standard* that

> Rose Atwood, a soiled dove of the bad lands, and Claudie Burrell, same section, were presented to the court on a charge of having imbibed too much tanglefoot and were reprimanded by the Squire with "My girls, please contribute $14.75 apiece to the city's funds and remember that the ways of the unjust are very rocky."

Those citations would be handed out by the court with predictable regularity, and the papers reported them in colorful language. Here's another one from May 1903:

> A lady (or at least she insisted she was one) sailing under the non-euphonious title of "Broken Nose Grace" was fined $5 and cost in police court Wednesday for conduct entirely unbecoming most people's idea of a lady.

Such reports had to assure the good ladies of Silverton that the city was doing its best to uphold the standards of decency and decorum while not at the same time forcing the town to raise its taxes.

At the beginning of a new year, the city streets had to be "purified." At least that's what the *Standard* reported in January 1904. Officers of the law had been doing that job. In the summer of 1902 they had "rounded up" and "moved out" of town "a mangy looking lot of people claiming to be Chilean gypsies..." In 1904, the paper could note with satisfaction, "a certain class with white hands and good clothes, who never disgrace themselves by earning a dollar, have been given notice to 'hike.' And a number of widows will be left on Blair Street." That way, we may suppose, each side's needs were taken care of.[22]

Yet Silverton's saddest days and ugliest incidents had occurred earlier in the new century. The chain of events had begun with rumors, reported in October 1901 in the *Herald* of neighboring Durango, that union laborers in Silverton were agitating to

expel the town's Chinese citizens. Meal prices in Durango and
in Silverton restaurants and mine boarding houses had been
raised with the support of organized labor to meet the general rise in living expenses, in particular those of eating house
employees. Chinese restaurants, however, were charging their
customary lower prices. That, the rumors had it, had prompted
Silverton's labor leaders to consider the expulsion of the Chinese.
They, however, promptly denied that any such move had been
contemplated. The rise in meal prices, they were reported to
have said, was "solely on account of the advance in prices of
provisions."[23]

It did not take long for the rumors reported in the *Durango
Herald* to take on concrete expression in anti-Chinese
sentiment reported in the Silverton papers. The *Silverton Miner*
did not hide its sympathy for what it called the "America for
Americans and white men" movement and its preference for
"all business in the camp to be conducted by white men." It
supported the unions' determination to see that "the undesir-
able citizens...dispose of their interest and get out of camp."
But only "legitimate means" should be used to achieve that
end, property should not be confiscated, and no one "engaged
in honest business" should be forced into financial ruin.[24]

The *Standard* was less circumspect. It printed with obvious
relish the statement of the Silverton Miners' Union that "whereas
the Chinese are a public nuisance and a detriment to the public
welfare" the union declared a boycott of all Chinese.

> As citizens they are a failure [declared the Union]; they do not
> assimilate as citizens and their habits are so obnoxious that they
> become the most undesirable class of people to the community
> in which they reside. The opium habit which has destroyed
> thousands of lives, is ready to destroy more if not checked. The
> Chinese dens in this city have destroyed over three hundred
> human beings...

Calling them "Chinks," the *Standard* reported that their jobs
as porters in the town's saloons had already been taken over
by "the colored gentry...and the one that can wear the highest collar, the loudest shirt and the flashiest clothes has the
greatest cinch on the job." A week later the paper added that

three-fourth of the "Chinks" had already left Silverton "and ere long the bland face of a Chinaman will be a curiosity." It then added, as a parting shot what it called a "poem:"

> And soon to all the outside world
> this message will be sent:
> We've whitened up old Silverton
> and the Chinaman 'has went.'[25]

It seems that not all "Chinamen had went" or that some of them had returned because the *Miner* reported that on Monday, May 12, "certain irresponsible parties" had in the middle of the night "committed theft in order to drive five Chinamen from town. Such acts of turbulent violence," added the *Miner*, "can but be condemned." The *Standard* gave a more detailed description of the events. A crowd of masked men had rounded up four or five Chinese, placed ropes around their necks and warned them to "hike" out of town. Shots had been fired at the "Spider's" Laundry on lower Blair Street, the O.K. Restaurant had been "cleaned out," and the fire bell was rung. The *Standard* further reported that one

> well known son of the flowered kingdom, was, on account of his many friendships, unmolested. Another celestial hid...and thus escaped a long walk. He was probably the one that rang the bell. It is understood that two of the poor wretches were quite roughly handled...while Policeman Dreyer was in another part of town.[26]

The *Standard's* chief concern, however, was not the damage done to the Chinese businessmen and people but the effect publicity of the events would have on Silverton's reputation. It informed its readers that Secretary of State John Hay had transmitted a request of the Chinese ambassador for protection "against threatened violence and forcible expulsion of Chinese at Silverton..." It printed an address to "perspective investors" in which it wrote that

> the hand full of miscreants who indulged in this provocation may be found in every town and city in the Union. While the STANDARD is in direct sympathy with any peaceable

movement to discourage Chinamen in their endeavor to occupy the positions and trades of American citizens, we believe there is a greater duty to perform in the ousting of such an element that resorts to such means as was employed at midnight hours.

It noted with approval that on Tuesday morning a citizen's committee had met with the leading Chinese, had assured them that they would not harass them and did not approve of violence, and had agreed with the Chinese on the "reasonable view" that "it was better for all concerned for them [the Chinese] to leave the city."[27]

In the following week the *Miner* approvingly related the organization of the General Welfare Association of San Juan County. This group, consisting of the town's leading businessmen and professionals, committed itself to the enforcement of all laws "in favor of, and against all persons alike, regardless of such person's race, color, nationality, occupation, or condition in life..." The group repeated once more that it was opposed to "the bringing of inferior races to our country," but it insisted that "when they do come and attempt to pursue a lawful occupation in a free land, they should be afforded the equal protection of the law." Mob law, the *Miner* added, is dangerous and Silverton must see to it that law and order prevails.[28]

And should there ever have been a lack of things to talk about, it was the weather that never failed to come to the rescue. The event of the decade was the storm of February 1905. It reminded old-timers of the blockade of 1884 when Silverton had been cut off from the world from February 4 to April 17 and ten or more feet of fallen snow and avalanches had blocked the Denver and Rio Grande and closed the mines of the district. This time the snow could not compare with that of 1884, but the winds were as fierce as they had been then. The storm broke loose on January 30 and was still blowing when the *Miner* reported it on February 17. The road to Durango was passable intermittently, and so was the way north to Ouray. But mail from the east could not get through over the Cumbres pass route. In Silverton snow accumulation had reached about four feet, though drifts had piled up to much greater heights. As in 1884, it was a storm to remember.[29]

CHAPTER 11

PUBLIC SCHOOLING'S TRIUMPH: HIGH SCHOOL GRADUATION

Silverton has cause to be proud of the record made by her public schools this year. The work has been most harmonious.... For the first time in its history the Silverton High school will hold commencement exercises. The class this year numbers four, Edith Dyson, Bertha Case, Jeanette Brown and Anna McNutt. These young ladies have completed the preparatory course, which admits them to the Freshman class of the state university, without an examination.... The exercises will take place at the Miners' Union Hall on June 20...

—*The Silverton Weekly Miner,* June 13, 1902

Silvertonians had reason to be proud of their public school on that first graduation day in the summer of 1902. During the preceding three school years enrollments had inched upward from 251 in 1899 to 253 in 1900 and 278 in 1901. Daily average attendance, though paralleling the rise in enrollments, had lagged behind with 174 in 1899, 217 in 1900, and 222 in 1901.

But even the enrollment figures were considerably lower than the school census numbers that counted as school children all persons between the ages of six and twenty-one who were by law eligible to attend public school. In 1899 those eligible amounted to 300 persons that included sixty-one between the ages of seventeen and twenty-one of which few were likely to

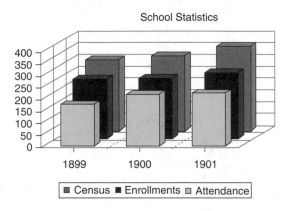

Figure 11.1   School Statistics 1899–1901

Source: School Reports in Silverton Newspapers and School Census Files in Silverton Public Library.

have attended school. In subsequent years the school census no longer listed the seventeen to twenty-one age-group separately, and counted 317 as the total number for 1900/1901 and 358 for 1901/1902.[1]

Translated into percentage figures we obtain the following distribution: in 1899/1900, 84 percent of all eligible persons were enrolled in Silverton's school, but only 69 percent of those enrolled attended with any regularity. In 1900/1901 the percentage of eligible persons who were enrolled had declined to 80 percent, but 86 percent of these attended school regularly. Finally, in the 1901/1902 school year the proportion of eligible persons who were enrolled had declined slightly to 78 percent, and 80 percent of these had attended regularly.

The 1899/1900 school year had opened with the eight elementary school grades and high school grades nine and ten, a principal and five teachers. The next year, when an eleventh grade was added, an assistant principal joined the staff and took over one of the five teaching positions. With the completion of the high school with a twelfth grade in 1901/1902 the school now had a principal, an assistant principal, and five teachers. In addition the district's budget listed teachers of the schools in Eureka and Howardsville.

Gradual, consistent growth had characterized Silverton's school during the first years of the new century and the first high school graduation had been its memorable high point. But had you asked Silvertonians how they felt about their school quite a few might have expressed uncertainty and bewilderment. They did not all agree that the time, money, and energy expended on the school's behalf had been worth the effort. As in earlier years, the newspapers printed repeated complaints over the negligence of parents who failed to send their children to school and over juvenile hoodlums and their "deviltries" on the streets. Compulsory education laws had not helped much as they exempted children living more than two miles from a school house, and the laws had been hard to enforce anyway.[2]

Those who kept a wary eye on the school district's budget were not pleased when early in 1900 the school directors floated the idea of erecting a new school building, one that would be designed specifically to accommodate the high school. When the May election of a new school board member drew nearer, the familiar conflict between the budget watchers and the school reformers broke out again. The *Standard* spoke out against possible bachelor candidates who "should stand aside for those who are directly interested," and against those who were motivated by "too much politics and by far too much personal prejudices..." The victims of these nay-sayers, the paper remarked, were in each case the teachers or the principal. All too often concerns for the welfare of the school were ignored "for the petit ambition of 'if you steal my cat, I'll steal your dog.' "[3]

The paper's concerns for teachers and principal were well grounded because when the school board announced the hires for the 1900/1901 school year, only one teacher, Miss Norah Greamba of the primary department, was a holdover from the previous year. All other positions, principal, assistant principal, high school assistant and two other teachers were newcomers. Principal Amsden did not seem to mind. He was ready to shift from teaching to speculating and prospecting. He put his money on the Magnet claim and seemed to think, so the *Standard*, that he had "the bonanza of the district in the Magnet." The paper was not happy with these developments. The board had not dealt fairly with teachers and principal. "A good deal of sharp talk before and since the school election has been going the rounds about...[last year's] teachers," and "there is always talk against the principal..." But "from our observation," wrote the paper, "and of the pupils themselves,...the faculty has done well."[4]

Board members, the new principal, Professor James A. Merriman, and the teachers were determined to make the new school year a success. The high school now comprised grades nine, ten, and eleven, carried an enrollment of seventeen students, and pursued for its curriculum the Latin Scientific Course, designed to prepare its students for college. Professor Merriman's first communication to the parents and patrons of the school, however, returned to a familiar complaint: Parental disinterest in school affairs. In September attendance had been as low as ninety-two out of an enrollment of 223. Perhaps it was illness or the threat of smallpox that had accounted for the low attendance, and in October the board ordered all children to be vaccinated.[5]

But other complaints soon turned up. "The larger and more advanced in deviltry kids" have principal Merriman "'buffaloed' and under a thorough state of subjection," wrote the *Standard*.

This wasn't exactly new, but it hurt when, as the *Standard* now wrote, the directors had fired one of Merriman's predecessors, Professor Vancil, in the summer of 1897 "simply because he didn't wear a silk hat and lift it to a certain element at every turn." And Vancil had been, in the paper's opinion, "one of the best school masters who ever had charge of our schools." Since then, the board had every year hired and then fired a new principal. Perhaps it was not accidental that a couple of weeks after the paper's complaint about Principal Merriman's inability to keep students under control, ex-principal and now successful mining entrepreneur Elmer E. Amsden advertised in the *Standard* that he was willing to "take a number of pupils who desire private lessons in any of the common school or high school branches."[6]

The *Standard's* slur against the school board's firing of Mr. Vancil was more than the board members could stand. Their treasurer Louis Wyman turned to the *Miner* and addressed "the Chronic School Kickers of Silverton:"

> We would be pleased to have you point out to us one school term at Silverton when some of you did not make yourselves as objectionable as possible, giving both the teachers and the school board all the trouble you could and we feel quite satisfied that if our dear Saviour came to earth again and tried to teach school in Silverton, he would make a failure of it, according to your way of thinking.
>
> What do you elect a school board for anyway? It is to look after the schools or simply to be kicked around by you for the fun of the thing? The writer has stood your abuse and nonsense for 18 months but proposes to stand it no longer....We propose to treat your children with respect and also propose to see that your children, as well as yourselves, treat our teachers with the same respect, otherwise we can have no such thing as a good school.
>
> Why don't you come and visit the schools instead of standing on the corners abusing everyone connected with them and spreading every rumor you may hear which is at all derogatory to the teachers or the school board?[7]

Alas, Mr. Wyman and his colleagues on the board had not heard the worst of all the bad news they might have feared. It came in February 1901 when the directors had to accept the resignations of three teachers, the Misses Norah Greamba, Mattie Walker, and Edna Short.[8] Unfortunately, we do not have the teachers' observations and reflections on what caused them to leave San Juan's Public School District. Was it that they, who taught the primary and intermediate grades, felt neglected and unappreciated by the three high school teachers, Principal Merriman, Assistant Principal Brown, and Miss Mary Lamont, the high school assistant? Was it that they were frustrated by the apparent inability of Principal Merriman to keep order in the school or by the school board members' hesitation or unwillingness to interfere? We do not know.

The *Standard* may give us some insight how at least some of Silverton's people thought about their school. Referring to "the late difficulty" the paper asked, "Whose fault is it?...Can't the

school board govern the school? Do they lack force to govern?"
It was not the fault of the teachers, the paper added.

> These ladies as regards teaching, gave perfect satisfaction to
> pupils and parents alike but the principal differed with them in
> many respects and so the conclusion arrived was that they could
> not dwell in harmony in the same building....If the school
> board had the courage that a real live school board should have
> had, [the paper continued,] they would have fired the principal
> and retained the teachers.

The paper then expressed its hope that the new teachers hired
to take the vacated places would "please all kickers in general."
The tragic death just a few days after her arrival in Silverton of
one of the newly hired teachers, Miss Frances Oaks, who was
to have taken over Miss Walker's grammar department, added
another somber note to the whole unhappy affair.[9]

Under its newly elected president, Josiah Watson, the school
board then was determined to prove to its critics that it was well
able to tackle all the school district's problems. It hired an attorney
to assist with any potential legal problems and it inaugurated a
public kindergarten for the summer of 1901 to be taught by
Mrs. Cora Harris. According to the *Standard*, Mrs. Harris
subsequently reported "unbounded success with her thirty-five
urchins." The board acquired property on Snowden Avenue to
provide proper space for the school's expanding primary department, and it elected a new principal, Mr. A.R. Lynch of near-by
Rico, and a high school assistant, Gertrude Upton. They were
to see to it that a full four-year high school course was to remain
a permanent part of Silverton's public school. Considering its
past difficulties with student absenteeism, the board decided to
add to the duties of the school's janitor those of a truant officer. His duty will be, wrote the *Standard*, "to see that no child
of school age remains away from school unless prevented by
illness or distance," and he "will have authority to take any
child away from employment and send him or her to school."
It is likely, added the *Miner*, "that the effort will be resisted in
some quarters, but it will be a good thing if properly enforced."
Once school had begun in September, the paper did not hesi-
tate to point out that enrollments were about 100 less than the

school census and that "there promises to be work enough to keep the truant officer busy."[10]

The board experienced difficulties, however, when it decided to open a school at Eureka to accommodate the settlement's fourteen pupils. The *Miner* reported the disheartening fact then when a young lady from Oklahoma, hired for the Eureka School, arrived in late October and witnessed first hand the situation in Eureka, she "decided to return to her former position and took the east-bound train the next morning." Troubled by this teacher's reaction, the board decided at its next meeting to have a new school built in Eureka and continued to search for a teacher. It finally succeeded late in November when Nora Colmer accepted the position.[11]

The school board's success in turning around the negative press it had received during past years showed in the supportive comments it received in December in both the *Miner* and the *Standard* when again unruly students and complaining parents vented their anger against teachers and school policy. The *Miner* praised school board members whose office, the paper wrote, "is a very thankless one and fraught with many an uncomplimentary remark from parents of the pessimistic order, whose motive seems to be the usurpation of the rules prescribed by both board and teacher." But, the *Standard* added, "the school board scored a victory, the matter was amicably settled, and the unruly children are back to school again." Both papers appealed to the town's parents to aid the school's authorities and to endorse any measures that the board felt necessary to take. The *Standard* subsequently noted with satisfaction that the town appeared to have heeded that advice. That "two youthful offenders" had been sent to the reform school "was evidence that the people are determined to break up the hoodlumism and general cussedness that has been so long prevalent in a portion of the Silverton youth."[12]

The papers' appeal found a ready response from the ladies of the San Juan Woman's Club who sponsored a mothers' club and invited parents and others concerned about school affairs to visit the school and observe its daily work. They called for a joint meeting of visitors and teachers to share views and become acquainted "to the end that a closer union may be brought

about between educators and patrons." The ladies of the San
Juan Woman's Club were concerned that mothers lost interest in their children's schooling when the latter reached the
intermediate grades, and the children suffered from that lack
of attention. Club members wanted to fill this lack through
"systematic school visitation" and asked mothers to join them
in that effort. A month later the *Standard* reported that "the
so-called mother's meeting…was more of a success than was
anticipated," and that at the next meeting the topic for dis-
cussion was to be: "How shall we secure regular attendance
of pupils throughout the year? More regular attendance of
pupils being the crying need of the Silverton schools…"[13] The
requests and complaints of past principals and school directors
had finally found a hearing.

The board also earned praise from the *Miner* at the beginning
of the new year when it let it be known that it considered the
construction of a new school house, though the paper had
reservations about a possible site and urged that a large play-
ground be included in the plans. Budgetary considerations then
postponed any action until 1905 when the board authorized a
six-room addition. It was not until the fall of 1911 that Silverton
received a new school house that is still in operation today.[14]

But the real triumph and the most visible achievement of San
Juan Public School District No. 1 was the graduation of its first
high school class on June 20, 1902. The class had asked the state
superintendent of public instruction, Mrs. Helen L. Grenfell,
to honor them with her presence in Silverton, and the city's
papers had done all they could to present the school, its board,
principal, teacher and students in the best possible light. "It
is the general opinion that this has been the best year in the
history of the Silverton schools," wrote the *Miner*. It attributed
this success to "the capability and business-like management of
the board and the efficient corps of teachers." The *Standard*
wrote of "peace and harmony between parent, teacher and
scholar.…Let the good work go on." The school board itself
underscored that assessment by unanimously re-electing the
entire teaching force for the new school year 1902/1903.[15]

Ever since the principalship of W.D. Cunningham in
the school year of 1894/1895 Silverton's school principals,

directors and many parents had dreamed of having their public school grow into a full-fledged high school of twelve grades. As had been repeatedly stated, such a school would employ a fully qualified college preparatory curriculum with the Latin Scientific Course that would allow its graduates to enter the freshman class of the state university without an examination. The other alternative, a high school that would prepare its students for entry into the world of work—preparing for life, as the phrase went, not preparing for college—was never talked about in Silverton.[16] The reason is not difficult to understand. Where the majority of the population sees and knows work as mining or as manual employment in support of mining, a high school education is not often thought of. And Silverton's middle-class professional and commercial parents had little incentive to seek a work preparatory high school education for their sons and daughters. For them a college preparatory education promised their children the future they aspired to.

As it was, Silverton's first graduation class did not include any boys. All its members were young ladies for whom their graduating class had in fact been a teacher preparation class, "a normal class," as it was called in many schools across the country, though never officially so in Silverton. Three of the four graduating seniors were soon to be employed in San Juan County schools. Bertha Katherine Case, a niece of teacher Mary Bertha Hodges, taught in the winter and spring of 1903 in the Gladstone school and in September in Howardsville. During the 1904/1905 school year she taught second grade in her alma mater, the Silverton school. To improve her chances at professional advancement she eventually enrolled in and graduated from the Colorado State Teachers College at Greeley in 1915.[17] Jeanette Brown taught at least one semester at the Chattanooga school. Anna McNutt was employed in the Eureka school and in 1904/1905 in the primary department at Silverton. The next year she moved up and taught Silverton's third grade. The fourth graduate, Edith Dyson, daughter of James Edgar Dyson, the former school board president, and of Alice Dyson, Silverton's teacher in 1880/1882, only taught school for a brief time at Rockwood, La Plata County, in the early 1920s. She subsequently attended Colorado State Agricultural College.

Silverton's high school graduation in 1902 had been a beginning. It did not matter that there had been no young gentlemen to send out into the world with a diploma. It did not matter that the young ladies did not immediately attend the State University or the State Teachers College. The high school offered these possibilities, and that is what mattered. The San Juan School District No. 1 was the immediate beneficiary of its high school. It had secured a supply of local teachers for its schools. Even though there were no graduates in the next class of 1903, a door had been opened, a way had been shown. Silverton was no longer a wild and far-off mining camp. It had a high school, and the future of its young who wanted to avail themselves of its offerings was assured.

POSTSCRIPT

LOOKING BACKWARD AND
LOOKING FORWARD

I have finished my story, the story of the first thirty years of a small American community and its school in the wilderness of the Rocky Mountains. As the twentieth century began this "metropolis of the San Juans," now no longer a hard-scrabble mining camp in Baker's Park, had become a town and proud county seat with a high school of its own that sent its children out into the world, well-prepared academically and morally to face the challenges that would await them there.

But what is the meaning of this all, what lessons can we gain from our story? To be sure, as I have written in the introduction, this story brings together chapters of the history of education, of women and of the West, and attempts thereby to let us gain a fuller, more rounded picture of what life was like in "them thare [*sic*] mountains." It tells how the lives of Silverton's denizens joined together and evolved into a living, breathing community whose members depended on each other for their survival.

It is a remarkable story of the gradual growth of community and of the role public education played in that process. Consider the male participants in this drama: Professionals of all kinds, businessmen, engineers, assayers, physicians, ministers, newspapermen, educators, hoteliers, bankers, lawmen; and rough-hewn outdoors men who worked as miners, road-builders, freighters, common laborers, blacksmiths, or rode in and out of town as never-do-wells and desperadoes. On the female side we have the wives of the middle-class professionals and of some of the miners as well as the "soiled doves" and their out-of-town sisters who came and left as they were tolerated in or chased out of town. And then there were the children, those "innocent cherubs" of middle-class families who, as the *San Juan Herald* described it, followed their "handsome petite schoolmarm...on their way down Reese Street,"[1] and those from the other side of the tracks whose "deviltries" annoyed the solid citizens and who, according to Elliot West, "panned for gold under sidewalks and in gutters in front of saloons and brothels, and [who] as they neared their teens...were often invited in."[2] Most of these folks, men, women and children, were of European stock, but the town also had its complement of "Chinks" and "colored gentry," as they were referred to.[3] They were alternately sought out for work or, in some instances of perverse racial discrimination, brutally expelled. Silverton's was not a classless society, and the potential for dissension and class warfare always lurked in the background and became acute during the 1890s with its nationwide labor strife and corporate capitalist exploitation.

How did Silverton avoid civil unrest and bloodshed? There were all along the threats of avalanches and house-high snow drifts in winter and of forest fires and floods in the summer. Surrounded as they were by the 11,000 to 13,000 feet high summits of Anvil, Tower, Kendall, Grand Turk, and Sultan Mountain and Storm, Macomber, and Little Giant Peak, the town's people saw daily how easily they could be cut off from the outside world by natural disasters. These threats made it clear to them that they depended on cooperation and mutual support for their survival and could not afford internal dissension.

But there was something else that held them together that did not depend on fear of outside threats of nature or of civil unrest in town and mines. And this was a consciousness of common destiny under any and all circumstance, whether under sunny or under cloudy skies. That consciousness arose from the determination to hold in balance individual liberty and social cohesion. Here Silverton's pioneer women played the central role, and their chosen instrument for achieving social stability was the public school. They could easily have turned to religion as well, as they were brought up as faithful members of religious congregations, and with their husbands held religious services at home. But religion would separate Silvertonians in their different denominations, and what was needed was not social separation but social cohesion. Thus with its classes for the children and its social gatherings, musicals, spelling bees, and public lectures the public school became the center of Silverton's educational and social life. It created community.

We see it in such endeavours as the successful campaign in 1886 for a public library that was meant to give the "boys" in the mines a place to relax other than the saloon around the corner. It didn't quite work that way as most of the readers came to be citizens of the town, but the intention is what matters. Far more successful was the involvement of Silverton's woman school pioneers in the politics of the town. Though women's suffrage became a reality in Colorado only in 1895—when it led to the election of the first woman state superintendent of public education—women had always been able to vote in Silverton school board elections, and had become strong advocates for citizen participation in elections on every level and for every office. They viewed voter participation in debates and voting as a key element in creating and sustaining community spirit and saw political parties as the chosen means to achieve that purpose.

So what is the meaning of the story I have told? It is our ability to view the creation of community in nineteenth century democratic America which we can observe in Silverton as in a raindrop that mirrors the wider universe of the United States: the slow growth of community from the bottom up built around and fostered by the public school as its central institution.

*   *   *

When that public school celebrated its One-hundredth Annual Commencement on May 22, 2003, it presented diplomas to four high school graduates, just as it had done at the same occasion in 1902. But this time one young man accompanied three young ladies. The commencement speaker noted the fact that the number of graduates in 2003 had been the same as that in 1902, but the school's enrollment figures were very different indeed. In September 1902 251 youngsters had crowded the classrooms of the school house that had been built in 1885, and a number of others attended school in Howardsville, Eureka, Gladstone, and Chattanooga. But in 2003, the splendid brick building that had been the pride of the town when it was built and opened in 1911, was the county's only public school for no more than sixty-two youngsters.

In that preceding century Silverton's citizens had witnessed the erection of a Carnegie Library in 1905 and in two-year intervals thereafter the construction of a county courthouse, a new town hall, and their beloved high school. In the years from 1910 to 1917 the town then experienced its greatest mining boom. Tragedy struck in 1918 with the worldwide flu epidemic. As Duane Smith reports, it killed approximately 150 people, at nearly 10 percent of the population, a higher percentage "than in any other town in the United States."[4]

During the depression years from 1925 to 1953 San Juan's mines were kept in operation, and World War II increased the demand for gold, silver, copper, lead, zinc, tungsten, and manganese. The fate of the town's commercial, professional, and blue-collar community and the school's enrollments reflected the ups and downs of the county's mining economy. As long as mining had flourished, family life had prospered, children

Figure 12.1   High School, Silverton, Colorado

crowded the school house, and the school board had been able to provide teachers for a full twelve grade curriculum. But when mining began to decline in the 1950s the picture darkened. To be sure, tourism began to gain a more prominent place in the local economy, but tourists did nothing to stop the drain on year-long residents, and the town's population shrank from 1,375 in 1950 to 890 in 1957, and to 716 in 1990.[5]

When then in 1991 the last producing hard rock mine, the Sunnyside, closed in August, disaster threatened Public School District No. 1. No longer did miners and prostitutes throng the town's saloons. No longer were stories told of fabulous strikes of gold and silver ore and gruesome death by avalanche and hypothermia. The town now struggled hard to turn itself into a tourist attraction. It sponsored brass band and folk music festivals, mining competitions in the summer, and ski races in the winter. But nothing could stop the steadily declining population. By 2000 it had dwindled to around 500.

For the school this meant a decline from 120 to 50 students and from a teaching staff of fourteen to six. It was no longer possible to continue offering separate classes as in the past when kindergarten and first grade, second and third grade, fourth and fifth grade, and each of sixth to twelfth grade had had their own homeroom teacher. What to do?

The answer came with the introduction in the fall of 2002 of the Expeditionary Learning—Outward Bound (ELOB) curriculum. The school was reorganized into four divisions, kindergarten to second grade, third to fifth grade, sixth to eighth grade, and the high school with eleven students in its combined ninth to twelfth grade. Six teachers and the principal directed learning in their classrooms and provided instruction in art, physical education, music, computers, and outdoor education.

To understand what are the defining characteristics of the ELOB curriculum and how it found its way at the outset of the twenty-first century to Silverton's public school in the Colorado Rocky Mountains we shall have to look for its origins a century ago in the Swabian Hills near Lake Constance. There it was first developed in *Schloss Schule Salem*, a coeducational boarding school, established by Prince Max of Baden in 1920. The school, a complex of three castles, was then, as it is now, surrounded by a landscape of picturesque beauty that, as historian George Mosse, one of the school's former students tells us, is diffused with legends of a past in which armed monks and knights loved their ladies and fought foreign invaders.[6]

When Prince Max of Baden opened his school in 1920 he appointed his friend and confidant, Kurt Hahn, to take on the duties of headmaster. Hahn, born in 1886 as son of a wealthy Jewish family in Berlin, had studied classical philosophy and philology at Oxford, Berlin, Heidelberg, Freiburg, and Göttingen. He had been employed during the war by the German Foreign Office to report on the British press and in 1919 to accompany the German delegation to the Versailles Peace Conference as advisor. At the *Schloss Schule Salem* Hahn developed and practiced his educational ideas that were designed to prepare German youth for a life of service to their communities and nation. As George Mosse experienced it, Hahn envisaged an elite of "soldierly" men who were to guide Germany out of its postwar chaos. Their pre-eminent traits of character were to be "a sense of duty, moral and physical courage, iron self-control, initiative, and compassion expressed through service to others."[7] Brian Simon, another historian and former pupil, did not share Hahn's concern "for an aristocratic education," but he admired his character and his never ceasing search for "the potential (artistic, intellectual, physical) which he believed existed in *every* pupil." Simon also endorsed Hahn's pedagogy designed to give "full rein and encouragement" to each student's potential.[8] That pedagogy reached far beyond an academic schooling of the mind. It included athletics and physical labor, training in arts and crafts, service to the community, and, above all, character formation as the capstone of all education.

Hahn served as Salem's headmaster from 1920 to 1933 when, as a courageous opponent of Hitler's brutalities, he was arrested and imprisoned by the Nazis. Through the help of British friends he was allowed to emigrate and in 1934 founded a boys' boarding school at Gordonstoun, Scotland, along the lines of his Salem school. Two years later, deeply concerned about the neglected physical education of English youth, he reached out beyond the students of his boarding school to neighboring working and school youths and included them in two-week long summer courses. Participants completing these received the Moray Badge that certified their training in physical education, swimming, two-day long expeditions in the country side, and instruction in rescue swimming and lifesaving.[9] The nearness of mountains and the Atlantic Ocean also prompted Hahn to include training for fire fighting, mountain rescue, and seamanship and to start a school coastguard service.[10]

When during wartime the Gordonstoun school was taken over by the British Army in 1940, Hahn and his students were evacuated to Llandinam, Wales. There Hahn ran a three-week long summer school course for students, apprentices, sea cadets, and soldiers. The Moray Badge was now replaced with a County Badge that was introduced in all of England's shires. To deserve the badge, a participant had to complete a fourfold program of training in physical education, including swimming, sprinting, distance running, jumping, and discus throwing; a study or performance project in some art or craft; an expedition to prove his mettle, and competence in some kind of service to others.[11] Later, in 1956, the County Badge was further promoted as the Duke of Edinburgh's Badge to be awarded throughout England and the British Commonwealth. To continue his seamanship training in his location in Wales Hahn rented a house at the shore near the village of Aberdovey, Cardigan Bay. In October 1941 he initiated there the four-week short courses that soon came to be called by the nautical term, Outward Bound.[12]

Hahn's educational ideas and practices reached the United States under the name of Outward Bound. American educators were motivated by the same concerns that had prompted Hahn at Gordonstoun to begin his short courses for young boys and men who were not necessarily students of his school: the deplorable physical conditions of youth; the unwillingness of conventional schooling to respond to youth's desire for adventure, and the absence of opportunities for youth to practice compassion and service. Upon learning of Outward Bound, Sargent Shriver, director of the Peace Corps, immediately ordered Outward Bound training to become part of the preparation for Peace Corps volunteers. The first Peace Corps Outward Bound school opened in 1961 at Arecibo, Puerto Rico, under the direction of William Sloane Coffin, the chaplain of Yale University.[13]

A year later, the first American Outward Bound school got underway on June 16, 1962, in the Rocky Mountains at Marble, Colorado. Here it was mountain rescue rather than sea rescue that stood at the center of its program. In the Superior-Quetico Boundary Wilderness of Minnesota a second American Outward Bound school broadened the program to year-round training that included women participants and canoeing expeditions as far north as Hudson Bay. At Hurricane Island near Maine's Penobscot Bay a school emphasized seamanship and navigation, including sea search and rescue. Other schools were established at North Carolina's Linville Gorge wilderness and Oregon's Three Sister Wilderness.[14]

Challenged by former Dean Paul Ylvisaker of the Harvard School of Education "to be a greater voice in education," Outward Bound set out to overcome its chief weakness, its lack of sustained engagement in a youngster's development due to its short-course nature. It now responded directly to the country's educational needs and deficiencies in its public schools. It applied its "experience-based education in mainstream and traditional schools using a mix of Outward Bound principles and the rigors of academic inquiry."[15] The result was the birth of Expeditionary Learning—Outward Bound (ELOB).

This is today one of the most promising and exciting ventures in school reform, a renewed introduction of principles of progressive education into mainstream public schools in the United States: the revival of the project method, the reliance on the students' abilities to begin a course of self-propelled lifelong learning, and, as demonstrated in learning expeditions and search and rescue as well as other aid activities, the intimate symbiosis of learning and real-life tasks.

By January of 1998 there were forty-seven ELOB schools in thirteen states, and by 2007 the number had risen to one-hundred-and-forty-seven schools in twenty-nine states, the District of Columbia, and Puerto Rico. On its Web site ELOB describes itself as

> a model for comprehensive school reform for elementary, middle and high schools. It emphasizes learning by doing, character growth, teamwork, and literacy. It connects academic learning

to adventure, service, work, and character development. It teaches teachers to teach reading, writing, science, math and other subjects through challenging sets of connected real-world projects called learning expeditions. Literacy instruction, embedded in learning expeditions, is a special focus of this design....It is a reapplication of Outward Bound's principles and practices to the business of everyday schooling.

The key to ELOB's design is the combination of commitment to high academic standards, to character as well as intellectual education, and the application of what has been learned to the service of others. Its design principles emphasize the role of challenge and adventure to rouse students' passions, to excite them to persevere and to use imagination, craftsmanship and self-discipline in the pursuit of tasks. They urge compassion for and service to others. They are based on Kurt Hahn's Seven Laws of Salem:

> Give children the opportunity for self-discovery; make the children meet with triumph and defeat; give children the opportunity of self-effacement in the common cause; provide periods of silence; train the imagination; make games (i.e.[,] competition) important but not predominant, and free the sons of the wealthy and powerful from the enervating sense of privilege.

The principles also incorporate the thoughts of Eleanor Duckworth's *The Having of Wonderful Ideas,* her book that encourages learning situations that build on children's curiosity and give children "time to experiment and time to make sense of what is observed." They also include Paul Ylvisaker's writings on intimacy and caring, and Theodore Sizer's observations on the responsibility for learning.[16]

How did ELOB succeed and emerge as one of the few path-breaking school reform proposals in the United States? The initiating step came with the requests for "break-the-mold" proposals for education reform sent out by the New American School Development Corporation (NAS) in the spring of 1991. A team pulled together by Outward Bound's Greg Farrell and Meg Campbell of the Chelsea, Massachusetts, School District responded and submitted a proposal of the ten design principles. The proposal turned out to be among the eleven NAS chose for funding from among the 690 submitted. As NAS' Liz Berry put it: when I first saw it, "my heart started pumping faster.... We had a loose charge: to take an idea of what an ideal school might look like, but make it realistic enough to put in place around the country."[17] And that is what they proceeded to do.

* * *

How, we must ask now, did Silverton come to adopt the ELOB curriculum and how has it fared with its choice? When during the 2001–2002 school year the school board was faced with the town's fiscal and economic woes described at the outset, it chose the Expeditionary curriculum as a promising way to rescue its school from further deterioration. By April of 2002 six of the school's teachers either were scheduled to retire or had to leave Silverton for other reasons. For the school year of 2002–2003 the board accepted a one-year exploratory contract with ELOB and started with a new group of teachers. A year later, with only one of the then employed six teachers resigning, faculty and board members agreed to continue the ELOB design as a long-term reform.

Since then the combination of an education of mind, heart, and hand with self-disciplined, conscientious service to others in the schools' multi-age classrooms have shown that this is the right education for Silverton's children. The school continues to practice Kurt Hahn's "experience therapy" and relies on physical education rather than team sports, emphasizes projects that demand careful preparation, employs expeditions that require perseverance, and highlights service that places the concern for one's own welfare under the commandment to love one's neighbor.[18] The school's strong tradition of outdoor adventure activities had made it easy to adopt ELOB projects and expeditions.

Silverton school's small population—in the years from 2003 to 2007 fluctuating between 56 and 74 students—has both its advantages and disadvantages. A visitor entering a classroom today is reminded of the small country school of yesteryear.

A small, multi-age classroom permits teachers to turn individualized instruction into the rule, and they can encourage and guide students to pursue their own research projects and connect these adventures to classroom learning. The mix of ages

Table 12.1 Silverton Public School. San Juan County School District #1 School Accountability Reports 2000–2007

Exploratory Full ELOB Contract Implementation in Effect of ELOB

	School Year 2000–2001	School Year 2001–2002	School Year 2002–2003	School Year 2003–2004	School Year 2004–2005	School Year 2005–2006	School Year 2006–2007
Elementary Performance: *Grades 1–5*	AVERAGE	LOW	LOW	AVERAGE	LOW	AVERAGE	AVERAGE
Academic Growth of Students	SIGNIFICANT DECLINE	STABLE	SIGNIFICANT IMPROVEMENT	SIGNIFICANT IMPROVEMENT	IMPROVEMENT	—	—
Middle School Performance: *Grades 6–8*	AVERAGE	AVERAGE	AVERAGE	LOW	LOW	AVERAGE	AVERAGE
Academic Growth Of Students	IMPROVEMENT	IMPROVEMENT	IMPROVEMENT	IMPROVEMENT	SIGNIFICANT DECLINE	IMPROVEMENT	—
High School Performance: *Grades 9–12*	AVERAGE	HIGH	AVERAGE	LOW	AVERAGE	AVERAGE	AVERAGE
Academic Growth of Students	SIGNIFICANT IMPROVEMENT	SIGNIFICANT DECLINE	SIGNIFICANT IMPROVEMENT	SIGNIFICANT DECLINE	IMPROVEMENT	—	—

Source: Adapted from www.cde.state.co.us.

also allows the older students to learn by teaching the younger ones. But small size also makes it difficult to comply meaningfully with the Colorado Student Assessment Program (CSAP) that calls for tests to rank the students' mastery of standards in reading, writing, mathematics, and science. This is so because for a school with very small classes the performance of one or two students "can effect the reported percentages for a particular level by anywhere from 20 to 100 percent in either a positive or negative direction," and thus distort the final results.[19]

Another evaluation measure that allows us to gain a sense of the school's success under the ELOB program is Colorado's annual School Accountability Reports (see table 12.1). They tell us that while in comparison with other schools in the state, the Silverton school in 2005–2006 ranked average, the students' academic growth in that year showed significant improvement in all grades, while in the 2006–2007 school year the results were mixed. The state also commended the district for having met all its accreditation requirements in both the 2004–2005 and the 2005–2006 school year. It found worthy of special mention the school's efforts on "communication within the community to focus on student achievement." The school itself had praised its community partners who had come forward "to offer music, art, foreign language, and computer classes to students in grades K-12."[20]

Under the *Leave No Child Behind* legislation the federal government, too, sets its targets for American schools and requires that a school district ensure that by the 2013–2014 school year every child in the nation is reading and performing mathematics at grade level. In the meantime each district annually has to reach state-set performance levels, known as Adequate Yearly Progress (AYP). The Silverton District has met this requirement ever since it was imposed in 2001.[21]

From the early 1870s to the first decade of the twenty-first century Silverton School District No. 1 has stood at the center of its community. The town's economic fortunes have had their ups and down, other institutions have come and gone, the school house remained. It assured the town, its citizens and its children, in good days and in bad, that this community was to prevail, proud of its past and confident of its future. Public education had given the town its identity, its reason for being, and its hope for the future.

SILVERTON PUBLIC SCHOOL STATISTICS

Term	Name	Office	Salary	Census of Children of School Age									Enrollments	
				6–21			6–16			16–21			Average	Attendance
				M	F	T	M	F	T	M	F	T		
Winter 1875/1876	Mrs. Matilda Huff	Teacher	—	18	15	33	—	—	—	—	—	—	24	15
Summer 1876	William K. Newcomb	Teacher	$ 75.00/month	22	18	40	—	—	—	—	—	—	23	15
Summer 1877	Miss Hulda Ann Puckett	Teacher	$ 75.00/month	24	18	42	—	—	—	—	—	—	20	—
Summer 1878	Mrs. Helen J. S. Bowman	Teacher	$ 64.65/month	36	18	54	—	—	—	—	—	—	33	30
Winter 1878/1879	Miss Emma Hollingsworth	Teacher	$ 64.65/month	22	25	47	—	—	—	—	—	—	—	—
Summer 1879	Eber C. Smith	Teacher	$ 87.50/month	—	—	44	—	—	—	—	—	—	38	24

Continued

Appendix I Continued

Term	Name	Office	Salary	Census of Children of School Age									Enrollments	
				6–21			6–16			16–21			Average Attendance	
				M	F	T	M	F	T	M	F	T		
Winter 1879/1880	James Edgar Dyson	Teacher	$ 87.50/month	30	28	58	23	20	43	7	8	15	32	22
Summer 1880	J. Homer Stewart	Teacher	$ 75.00/month	—	—	—	—	—	—	—	—	—	—	—
Winter 1880/1881	Mrs. James Edgar Dyson	Teacher	$ 75.00/month	24	30	54	—	—	—	—	—	—	36	22
Summer 1881	William K. Newcomb	Teacher	$ 75.00/month $ 225.00/term	—	—	—	—	—	—	—	—	—	—	—
Year 1881/1882	Mrs. James Edgar Dyson	Teacher	$ 88.00/month $ 276.25/term $ 800.00/year	24	30	54	16	25	41	8	5	13	36	22
Year 1882/1883	Miss Isabella Munroe	Teacher	$ 88.00/month $266.66/term $ 800.00/year	36	36	72	29	26	55	42	29	71	39–42	33
Year 1883/1884	Miss Isabella Munroe	Principal Grammar Dep.	$ 800.00/year	75	82	157	59	69	128	16	13	29	75	37
	Miss Mary E. Gaines	Ast. Teacher	$ 674.50/year	—	—	—	—	—	—	—	—	—	—	—
Year 1884/1885	Miss Isabella Munroe	Intermed. Dep.	$ 800.00/year	81	71	152	49	51	100	32	20	52	85	25
	Miss Armes	Primary Dep.	$ 75.00/month	—	—	—	—	—	—	—	—	—	—	—

Year	Name	Position	Salary											
	Miss Armes	Intermed. Dep.	$75.00/month	—	—	—	—	—	—	—	—	—	—	—
	Miss Cora Downer	Primary Dep.	$75.00/month	—	—	—	—	—	—	—	—	—	—	—
Year 1885/1886	Miss Cora Downer	Principal	$85.00/month	99	84	183	50	62	112	49	22	71	80	—
		Intermed. Dep	—	—	—	—	—	—	—	—	—	—	—	—
	Miss Frances Tracy	Primary Dep.	$75.00/month	—	—	—	—	—	—	—	—	—	—	—
Year 1886/1887	Mrs. E. W. Hodges	Principal	$35.00/month	86	97	183	62	78	140	24	19	43	105	70
	Miss Susie Robin	Ast. Principal	$75.00/month	—	—	—	—	—	—	—	—	—	—	—
Year 1887/1888	Prof. C. M. Kiggins	Principal	$133.00/month	59	88	147	44	82	126	15	6	21	91	—
	Mrs. U. Tom Garrett	Principal	$90.00/month	—	—	—	—	—	—	—	—	—	—	—
	Miss Susie Robin	Primary Dep.	$80.00/month	—	—	—	—	—	—	—	—	—	—	—
Year 1888/1889	Frank Prentiss	Principal	replaced by	56	69	125	43	64	107	13	5	18	91	89
	Mrs. U. Tom Garrett	Principal	$90.00/month	—	—	—	—	—	—	—	—	—	—	—
	Miss Luella Burgwin	Ast. Principal	$90.00/month	—	—	—	—	—	—	—	—	—	—	—
		Primary Dep.	—	—	—	—	—	—	—	—	—	—	—	—
	Miss Mary Burgwin	Ast. Teacher Upper Grade	$90.00/month	—	—	—	—	—	—	—	—	—	—	—

Continued

Appendix I Continued

Term	Name	Office	Salary	Census of Children of School Age 6–21 M	F	T	6–16 M	F	T	16–21 M	F	T	Enrollments	Average Attendance
Year 1889/1890	Miss Mary Burgwin	Principal / Upper Grade	$ 95.00/month	71	86	157	50	68	118	21	18	39	103	52
	Miss Luella Burgwin	Primary Grade	$ 95.00/month	—	—	—	—	—	—	—	—	—	—	—
Year 1890/1891	Miss Mary Burgwin	Principal / Upper Grade	$ 100.00/month	82	71	153	70	58	128	12	13	25	109	—
	Miss Luella Burgwin	Lower Grade	$ 100.00/month	—	—	—	—	—	—	—	—	—	—	—
	Miss Mary Roth	Ast. Teacher / Lower Grade	—	—	—	—	—	—	—	—	—	—	—	—
Year 1891/1892	John T. Barnett	Principal	$ 100.00/month	89	101	190	77	71	148	12	30	42	133	—
	Miss Mary Ross	Primary Dep.	$ 100.00/month	—	—	—	—	—	—	—	—	—	—	—
	Mrs. U. Tom Garrett	Primary Dep.	$ 100.00/month	—	—	—	—	—	—	—	—	—	—	—
	Miss M. Capwell	Intermed. Dep.	—	—	—	—	—	—	—	—	—	—	—	—
Year 1892/1893	John T. Barnett	Principal	$ 100.00/month	101	101	202	84	79	163	17	22	39	116	—

Name	Department	Salary											
Miss Louise Walling	Intermed. Dep.	$ 75.00/month	—	—	—	—	—	—	—	—	—	—	—
Miss Luella M. Liscomb	Primary Dep.	$ 75.00/month	—	—	—	—	—	—	—	—	—	—	—
Mrs. J. S. Robinson		—	—	—	—	—	—	—	—	—	—	—	—
Summer School													
Year 1893/1894 John T. Barnett	Principal	$ 120.00/month	84	105	189	68	70	138	16	34	50	—	—
Miss Mae Jones	Intermed. Dep.	$ 35.00/month	—	—	—	—	—	—	—	—	—	—	—
Mrs. Mary B. Hodges	Primary Dep.	$ 35.00/month	—	—	—	—	—	—	—	—	—	—	—
Year 1894/1895 W. D. Cunningham	Principal	$ 120.00/month	81	93	174	72	67	139	10	25	35	150	125
Miss Mae Jones	Intermed. Dep.	$ 85.00/month	—	—	—	—	—	—	—	—	—	—	—
Mrs. Mary B. Hodges	Primary Dep.	$ 85.00/month	—	—	—	—	—	—	—	—	—	—	—
Year 1895/1896 F. M. Vancil	Principal	$ 120.00/month	91	101	192	80	73	153	11	28	39	185	154
Sadie Knowlton/ Burnsides	Intermed. Dep.	$ 89.25/month	—	—	—	—	—	—	—	—	—	—	—
Mrs. Jennie M. Reavley	Grammar Dep.	$ 89.25/month	—	—	—	—	—	—	—	—	—	—	—
Mrs. Mary B. Hodges	Primary Dp.	$ 89.25/month	—	—	—	—	—	—	—	—	—	—	—
Beginning of High School													
Year 1896/1897 F. M. Vancil	Principal	$ 1077.00/ year	103	132	235	—	—	—	—	—	—	209	125–211

Continued

Appendix I Continued

Term	Name	Office	Salary	6–21			6–16			16–21			Enrollments	
				M	F	T	M	F	T	M	F	T		Average Attendance
	Miss Mable Daniels	Ast, Principal	$ 756.50/year	—	—	—	—	—	—	—	—	—	—	—
	Mrs. Jennie Reavley	Grammar Dep.	$ 750.00/year	—	—	—	—	—	—	—	—	—	—	—
	Mrs. Mattie Carpenter	Intermed. Dep.	$ 490.00/year replaced in April	—	—	—	—	—	—	—	—	—	—	—
	Mrs. Lefa D. Clark	Intermed. Dep.	$ 300.00/year	—	—	—	—	—	—	—	—	—	—	—
	Mrs. Mary B. Hodges	Primary Dep.	$ 850.00/year	—	—	—	—	—	—	—	—	—	—	—
Year 1897/1898	F. E. Mullen	Principal	$ 1000.00/year	133	147	280	114	108	122	19	29	48	238	92–220
	Miss Mable Daniels	Ast. Principal	$ 80.00/month	—	—	—	—	—	—	—	—	—	—	—
	Miss Rose Lee Smith	Teacher	$ 80.00/month	—	—	—	—	—	—	—	—	—	—	—
		Librarian	$ 5.00/month	—	—	—	—	—	—	—	—	—	—	—
	Miss Nellie O'Donoghue	Intermed. Dep.	$ 80.00/month	—	—	—	—	—	—	—	—	—	—	—
	Miss Cora Edwards	Primary Dep.	$ 80.00/month	—	—	—	—	—	—	—	—	—	—	—
Year 1898/1899	Prof. Edward L. Howett	Principal	$ 125.00/month	140	152	292	121	117	238	19	35	54	260	170

Miss Virginia McMechen	Ast. Principal	$ 80.00/ month	—	—	—	—	—	—	—	—	—	—	—
Miss Rose Lee Smith	Grammar Dep.	$ 80.00/ month	—	—	—	—	—	—	—	—	—	—	—
	Librarian	$ 5.00/month	—	—	—	—	—	—	—	—	—	—	—
Miss Nellie O'Donoghue	Intermed. Dep.	$ 80.00/ month	—	—	—	—	—	—	—	—	—	—	—
Miss Cora Edwards	Primary Dep.	$ 80.00/ month	—	—	—	—	—	—	—	—	—	—	—
Year 1899/1900													
Elmer E. Amsden	Principal	$ 900.00/year	150	150	300	118	121	239	32	29	61	251	174
Miss Virginia McMechen	Ast. Principal	$ 90.00/ month	—	—	—	—	—	—	—	—	—	—	—
Miss Rose Lee Smith	Grammar Dep.	$ 80.00/ month	—	—	—	—	—	—	—	—	—	—	—
	Librarian	$ 5.00/month	—	—	—	—	—	—	—	—	—	—	—
Miss Nellie O'Donoghue	Intermed. Dep.	$ 80.00/ month	—	—	—	—	—	—	—	—	—	—	—
Miss Nora Greamba	Primary Dep.	$ 80.00/ month	—	—	—	—	—	—	—	—	—	—	—
Year 1900/1901													
Prof. James A. Merryman	Principal	$ 125.00/ month	157	160	317	—	—	—	—	—	—	253	217–223
Prof. L. E. Brown	Ast. Principal	$ 100.00/ month	—	—	—	—	—	—	—	—	—	—	—
Miss Mary Lamont	High School Ast.	$ 80.00/ month	—	—	—	—	—	—	—	—	—	—	—
Miss Mattie Walker	Grammar Dep.	$ 80.00/ month replaced in February	—	—	—	—	—	—	—	—	—	—	—
Miss Nellie Donoghue	Grammar Dep.	$ 85.00/ month	—	—	—	—	—	—	—	—	—	—	—

Continued

Appendix I Continued

Term	Name	Office	Salary	Census of Children of School Age									Enrollments	
				6–21			6–16			16–21				
				M	F	T	M	F	T	M	F	T		Average Attendance
	Miss Edna Short	Intermed. Dep.	$ 80.00/ month replaced in February	—	—	—	—	—	—	—	—	—	—	—
	Miss France Oakes	Intermed. Dep.	$ 80.00/ month replaced in March	—	—	—	—	—	—	—	—	—	—	—
	Miss I. Coda Holman	Intermed. Dep.	$ 80.00/ month	—	—	—	—	—	—	—	—	—	—	—
	Miss Nora Greamba	Primary Dep.	$ 80.00/ month replaced in February	—	—	—	—	—	—	—	—	—	—	—
	Miss M. J. Ableson	Primary Dep.	$ 80.00/ month	—	—	—	—	—	—	—	—	—	—	—
Summer 1901	Mrs. Cora Harris	Kindergarten	$ 60.00/ month	—	—	—	—	—	—	—	—	—	—	—
Year 1901/1902	Andrew R. Lynch	Principal and Superintendent	$ 125.00/ month	184	174	358	—	—	—	—	—	—	278	209–223
	J. B. McDonough	Ast. Principal	$ 100.00/ month Temporarily replaced	—	—	—	—	—	—	—	—	—	—	—
	Mrs. E. W. Hodges	2nd Primary	$ 85.00/ month	—	—	—	—	—	—	—	—	—	—	—
	Mr. F. A. Swayne	Ast. Principal	$ 80.00/ month	—	—	—	—	—	—	—	—	—	—	—

Miss Gertrude Upton	High School Ast.	$ 80.00/month	—	—	—	—	—	—	—	—	—	—	—	—
Mrs. Lena Carr	Grades 4 and 5 Librarian	$ 85.00/month	—	—	—	—	—	—	—	—	—	—	—	—
Miss Coda Holman	Grades 2 and 3	$ 80.00/month	—	—	—	—	—	—	—	—	—	—	—	—
Miss M. J. Ableson	Primary Dep.	$ 80.00/month	—	—	—	—	—	—	—	—	—	—	—	—

Graduates of First High School Commencement

Miss Anna C. Nutt
Miss Jeanette H. Brown
Miss Bertha K. Case Miss Miss Edith M. Dyson

Red Mountain Summer School

Summer 1887	Miss Kittie Real	—	—	—	—	—	—	—	—	—	—	—	—	—
Summer 1891	Miss Luella Burgwin	—	—	—	—	—	—	—	—	—	—	—	—	—
Summer 1893	Mrs. J. S. Robinson	—	—	—	—	—	—	—	—	—	—	—	—	—

Howardsville School

Year 1897/1898	Miss Alice Marian Henderson	Teacher	$ 35.00/month plus board	—	—	—	—	—	—	—	—	—	—	—
Summer 1898	Miss Hedrickson	Teacher	$ 55.00/month	—	—	—	—	—	—	—	—	—	—	—
	Miss Cora Edwards	Kindergarten	$ 50.00/month	—	—	—	—	—	—	—	—	—	—	—

Continued

Appendix I Continued

Term	Name	Office	Salary	Census of Children of School Age													Enrollments	
				6–21			6–16			16–21							Average Attendance	
				M	F	T	M	F	T	M	F	T						
Year 1898/1899	Miss Hendrickson	Teacher	$ 55.00/month	—	—	—	—	—	—	—	—	—					—	—
Summer 1899	Miss Annie Gray	Teacher	$ 55.00/month	—	—	—	—	—	—	—	—	—					—	—
Year 1899/1900	Miss Annie Gray	Teacher	$ 55.00/month	—	—	—	—	—	—	—	—	—					—	—
Summer 1900	Miss Louise Pyle	Teacher	$ 55.00/month	—	—	—	—	—	—	—	—	—					—	—
Summer 1901	Miss Jessie Munyon	Teacher	$ 55.00/month	—	—	—	—	—	—	—	—	—					—	—
Year 1901/1902	Miss Jessie Munyon	Teacher	$ 55.00/month	—	—	—	—	—	—	—	—	—					—	—
Eureka School																		
Year 1901/1902	Miss Louise Pyle	Teacher	$ 55.00/month replaced by	—	—	—	—	—	—	—	—	—					—	—
	Miss Nora Colmer	Teacher	$ 55.00/month	—	—	—	—	—	—	—	—	—					—	—

Notes:
M: Male
F: Female
T: Total

Appendix II

School Directors
of Silverton Public
School District, 1877–1906

May 1877	George Walz
	William Edward Earl
	John L. Pennington
May 8, 1880	President: Dr. Robert H. Brown
	Secretary: John M. Rogers
	Treasurer: Leander F. Hollingsworth
May 24, 1881	President: John M. Rogers
	Secretary: Dr. Robert H. Brown
	Treasurer: William Edward Earl
May 1, 1882	President: John M. Rogers
	Secretary: Leander F. Hollingsworth
	Treasurer: William Edward Earl
December 18, 1882	President: John M. Rogers
	Secretary: Harlan Page Roberts
	Treasurer: W. Edward Earl
May 7, 1883	President: James Edgar Dyson
	Secretary: Harlan Page Roberts
	Treasurer: William Edward Earl
May 6, 1884	President: James Edgar Dyson
	Secretary: Harlan Page Roberts
	Treasurer: C. M. Frazier
September 2, 1884	President: James Edgar Dyson
	Secretary: John Montgomery
	Treasurer: C. M. Frazier

May 4, 1885	President: James Edgar Dyson Secretary: Jonathan W. Fleming Treasurer: C. M. Frazier
August 29, 1885	President: Nathaniel Ellmaker Slaymaker Secretary: Jonathan W. Fleming Treasurer: C. M. Frazier
May 3, 1886	President: Mrs. John (Maggie Williamson) Montgomery Secretary: Jonathan M. Fleming Treasurer: C. M. Frazier
May 5, 1887	President: Mrs. John (Maggie Williamson) Montgomery Secretary: W. J. Forsyth Treasurer: David Ramsey
November 17, 1887	President: Mrs. John (Maggie Williamson) Montgomery Secretary: W. J. Forsyth Treasurer: Benjamin Austin Taft
May 7, 1888	President: W. Lafayette "Lafe" Henry Secretary: C. M. Frazier Treasurer: Thomas Andrew Gifford
May 6, 1889	President: Robert J. Bruns Secretary: C. M. Frazier Treasurer: Thomas Andrew Gifford
May 10, 1890	President: Robert J. Bruns Secretary: C. M. Frazier Treasurer: Horace Greely Prosser
June 1, 1891	President: Robert J. Bruns Secretary: Fred G. Helmboldt Treasurer: Thomas Andrew Gifford
May 3, 1892	President: J. Frank Molique Secretary: Fred G. Helmboldt Treasurer: Thomas Andrew Gifford
May 2, 1893	President: J. Frank Molique Secretary: Fred G. Helmboldt Treasurer: William Lyle
June 12, 1893	President: J. Frank Molique Secretary: Mart Stockman Treasurer: William Lyle
May 7, 1894	President: J. Frank Molique Secretary: John Rogers Treasurer: William Lyle

February 4, 1895	President: J. Frank Molique
	Secretary: John Rogers
	Treasurer: Mrs. M. B. Starkweather
May 6, 1895	President: Samuel Uriah Morris
	Secretary: John Rogers
	Treasurer: D. W. Williams
May 14, 1896	President: Samuel Uriah Morris
	Secretary: John Rogers
	reasurer: Daemerit "Dee" Umbell
May 3, 1897	President: Samuel Uriah Morris
	Secretary: Charles E. Robin
	reasurer: Daemerit "Dee" Umbell
May 2, 1898	President: Julius Brice Patterson
	Secretary: Charles E. Robin
	Treasurer: Daemerit "Dee" Umbell
May 2, 1899	President: Julius Brice Patterson
	Secretary: Charles E. Robin
	Treasurer: Louis George Wyman
May 7, 1900	President: Julius Brice Patterson
	Secretary: Fred Goble
	Treasurer: Louis George Wyman
May 10, 1901	President: Josiah Watson
	Secretary: Fred Goble
	Treasurer: Louis George Wyman
May 5, 1902	President: Josiah Watson
	Secretary: Fred Goble
	Treasurer: Louis George Wyman
May 4, 1903	President: Josiah Watson
	Secretary: Frank L. Palmer
	Treasurer: Louis George Wyman
May 4, 1904	President: William Marsh
	Secretary: Frank L. Palmer
	Treasurer: Louis George Wyman
February 18, 1905	President: William Marsh
	Secretary: S. Dave Cunningham
	Treasurer: Louis George Wyman
February 23, 1905	President: B. B. Allen
	Secretary: S. Dave Cunningham
	Treasurer: Louis George Wyman
May 1, 1905	President: S. Dave Cunningham
	Secretary: Charles E. Robin
	Treasurer: Jesse S. Baily

May 6, 1905   President: S. Dave Cunningham
               Secretary: Charles E. Robin
               Treasurer: Jesse S. Baily
May 7, 1906   President: S. Dave Cunningham
               Secretary: Charles E. Robin
               Treasurer: Jesse S. Baily

Appendix III

San Juan County
Superintendents of Public Schools

State of Colorado: La Plata County

1874–1877   Jacob M. Hanks

State of Colorado: San Juan County

1877–1878   Elected October 3, 1876, William Munroe (R)
1878–1880   Elected Fall 1877, William Munroe (R)
1880–1882   Henry O. Montague, Silverton (R)
1882–1884   Dr. med. Robert H. Brown, Silverton (D)
1884–1886   Dr. med. Robert H. Brown, Silverton (D)
1886–1888   Dr. med. J. N. Pascoe, Silverton (R)
1888–1890   Dr. med. J. W. Brown, Silverton (D)
1890–1892   Dr. med. J. W. Brown, Silverton (D)
1892–1894   Dr. med. J. N. Pascoe, M.D. (R)
1894–1896   Dr. med. J. N. Pascoe, M.D. (R)
1896–1898   Mrs. Ellen Carbis (P)
1898–1900   Mrs. Ellen Carbis (P)
1900–1902   Mrs. Ellen Carbis (P)
1902–1904   Mrs. Ellen Carbis (P)

Appendix IV

Colorado Territorial Superintendents of Education

1861–1863 W. J. Curtice, appointed by Governor Gilpin
1863–1865 W. S. Walker, appointed by Governor Evans
1865–1866 F. W. Atkins, ex-officio as treasurer
1866–1867 F. W. Atkins, ex-officio as treasurer
1867–1869 Columbus Nuckolls, ex-officio as treasurer
1870–1872 Wilbur C. Lothrop, appointed by Governor McCook
1872–1873 Wilbur C. Lothrop, reappointed by Governor McCook
1873–1874 Horace M. Hale, appointed by Governor Elbert
1874–1876 Horace M. Hale, reappointed by Governor Elbert

State Superintendents of Education

1876–1878 Jos. C. Shattuck, elected (R)
1878–1880 Jos. C. Shattuck, reelected (R)
1880–1882 Leonidas S. Cornell, elected (R)
1882–1884 Jos. C. Shattuck, elected (R)
1884–1886 Leonidas S. Cornell, elected (R)
1886–1888 Leonidas S. Cornell, reelected (R)
1888–1890 Fred Dick, elected (R)
1890–1892 Nathan B. Coy, elected (D)
1894–1896 Mrs. Agnette J. Peavey, elected (R)
1897–1899 Miss Grace Espy Patton, elected (D)
1898–1900 Mrs. Helen Loring Grenfell, elected (Silver Republican)
1900–1902 Mrs. Helen Loring Grenfell, elected
1902–1904 Mrs. Helen Loring Grenfell, elected

Appendix V

Silverton Public Schools, Rules for Pupils

1. No pupil affected with any contagious disease, or coming from a house in which such disease exists, shall be allowed to remain in the public school.
2. Any child coming to school without proper attention having been given to the cleanliness of his person or dress, or whose clothes need repairing, shall be sent home to be properly prepared for the school-room.
3. No pupil shall be allowed to retain connection with the public school unless furnished with books, slate and other utensils required; provided, that no pupil shall be excluded for such cause, unless the parent or guardian shall have a week's notice and be furnished by the teacher by a list of the books or articles needed, and then with proper notice the school board may furnish books as provided in section 51, ninth, of the school law of Colorado.
4. Excuses for tardiness, absence or dismissal before the close of school must be made in writing by the parent or guardian.
5. Six half-days absence in any four consecutive weeks—sickness alone accepted—shall render pupil liable to suspension.
6. Pupils must leave the school premises and go directly home after school is closed, both at noon and night, unless otherwise permitted by the teacher and shall be subject to the rules of the school during their passage each way, and must not bring to the school, books or papers foreign to the purpose of study.

7. Pupils are forbidden to throw stones or missiles of any kind upon the school grounds or in the streets in the immediate vicinity of the school grounds.

8. Pupils shall not mark, scratch or break in any way the books, furniture, casings, walls, windows, fences or any of the appurtenances of the school premises. Pupils committing such injuries, accidental or intentional, shall immediately procure the necessary repair or be assessed by the school board a sum sufficient to cover the damage; and on refusal to comply with this rule may be expelled from school.

9. The use of tobacco or profane or other improper language, indecent exposure and fighting are strictly prohibited.

10. Any pupil that may be aggrieved or wronged by another pupil may report the fact to his teacher. No pupil in any case shall attempt to avenge his own wrong.

11. The promotion of pupils from one grade to another shall be made at such times as the interest of the school may require.
    Pupils may be sent into the grade next below the grade to which they belong whenever their scholarship falls below the standard fixed for admission to the grade by the principal,
    but such pupils may be permitted to regain their lost position within one month if their scholarship warrants it.

12. Pupils must not arrive at the school building sooner than thirty minutes before 9 o'clock in the forenoon, or fifteen minutes before 1 o'clock in the afternoon. The doors will not be opened earlier.

13. Pupils shall walk quietly in and out from their school rooms through the hall; make no noise at any time in any part of the building.

Adopted by the School Board February 13, 1889

SILVERTON PUBLIC SCHOOLS, RULES FOR TEACHERS

1. Teachers shall be at their school-rooms twenty minutes before the opening of school in the morning and fifteen minutes in the afternoon.
2. It shall be the duty of the teachers to practice such discipline in their school as would be exercised by a kind and judicious parent in his family, always firm and vigilant, but prudent. They shall endeavor on all proper occasions to impress upon the minds of their pupils the principles of morality and virtue, a sacred regard for truth, neatness, sobriety, industry and frugality.
3. Teachers shall insist on obedience to school rules of pupils not their own when the teachers of such pupils are not at hand.
4. The hours of tuition shall be from 9 o'clock in the forenoon to 11:45—without recess; and from 1:15 o'clock in the afternoon to 4 p.m.—without recess.
5. The opening and closing exercises of the school shall be conducted as the teacher may deem advisable.
6. The principal shall have the power to suspend from the privileges of the school, pupils guilty of gross misconduct or contin-ued insubordination to school regulations. The principal shall present to the board at their next regular meeting full particulars concerning each such case of discipline.
7. Each teacher is required to have a copy of the school regulations in his or her school room and to read to the scholars at least once in each term so much of the same as will give them a just

understanding of the rules which apply to them and by which they are governed.

8. In cases not provided for under these rules the principal shall employ such temporary measures as are necessary to maintain good order in the school and preserve their efficiency.

9. Teachers are instructed to enforce the rules and regulations adopted by the school board and will have their support and cooperation.

10. Section 51, seventh, says that the school board shall have power, and it shall be their duty, to suspend or expel pupils from school who refuse to obey the rules thereof.

Adopted by the School Board, February 13, 1889

NOTES

INTRODUCTION

1. Jürgen Oelkers, *Reformpädagogik: Eine kritische Dogmengeschichte* (Weinheim und München, Germany: Juventa, 1989); see also my "Toward a Theory of Progressive Education?" *History of Education Quarterly*, 37, 1 (Spring 1997), 45–59.
2. Allen Nossaman, *Many More Mountains*, 3 vols. (Denver, CO: Sundance Publications, 1989–1998).
3. Freda Carley Peterson, *The Story of Hillside Cemetery, Silverton, San Juan County, Colorado*, 2 vols. (Silverton, CO: F.C. Peterson, 1996–1998).

1 BEGINNINGS

1. Colorado passed compulsory school attendance legislation in 1889.
2. Territory of Colorado, *Third Biennial Report of the Superintendent of Public Instruction for the Two Years Ending September 30th, 1875* (Denver, CO: Rocky Mountain News Steam Printing House, 1876), 98. Subsequent summary statistics issued by the State Superintendent reduced the number to 24 students, twelve boys and twelve girls.
3. Allen Nossaman, *Many More Mountains*, vol. 1: *Silverton's Roots* (Denver, CO: Sundance Publications, 1989), 201, 229, 282, 299, and Ibid., vol. 2: *Ruts into Silverton* (Denver, CO: Sundance Publications, 1993), 11.
4. Nossaman, 1:228–230, 249–250, 289, 309, 323; 2:43.
5. Ibid., 1:276.
6. Ibid., 1:97–147.
7. Ibid., 1:103, 109, 115, 123, 128, 135; 2:11; Lena S. Knapp, "John W. Wingate (Patriot)" in Sarah Platt Decker Chapter D.A.R., *Pioneers of the San Juan Country* (Colorado Springs, CO: Out

West Printing and Stationary Co., 1952), 3:149–150, and Hans Aspaas, "The Aspaas Family," in Platt Decker Chapter D.A.R., *Pioneers of the San Juan Country*, 3: 159.
8. Nossaman, 1:134, 141–145, 158, 204, 294–295.
9. Ibid., 1:275.
10. Ibid., 1:222, and Freda Carley Peterson, *The Story of Hillside Cemetery, Silverton, San Juan County, Colorado* (Silverton, CO: F.C. Peterson, 1996), 1:H-15.
11. Peterson, 1:H-17.
12. Nossamann, 1:223, and Aspaas, "The Aspass Family," in Platt Decker Chapter D.A.R., *Pioneers of the San Juan Country*, 3:159.
13. Peterson, 1:C-50, L-22; 2:N-14.
14. Nossaman, 1:203, 204, 223, 225; *San Juan Herald*, September 8, 1881.
15. Nossaman, 1:205, 225; Aspaas, "The Aspass Family," in Platt Decker Chapter D.A.R., *Pioneers of the San Juan Country*, 3:159.
16. See the 1860–1882 "Chronological History" in the *La Plata Miner* of December 30, 1882 and Nossaman, 1:201, 276, 302.
17. Territory of Colorado, *Third Biennial Report*, 98.
18. However, Ruth Rathmell, author of *Of Record and Reminiscence—Ouray and Silverton* (Ouray, CO: *The Ouray County Plaindealer* and *Herald*, 1977) wrote in the *Silverton Standard and* the *Miner* of March 14, 1975, that Will K. Newcomb was the teacher when the school opened.

2   A House of Many Uses

1. See the *First Biennial Report of the Superintendent of Public Instruction of the Territory of Colorado for the School Year Ending September 30, 1870 and September 30, 1871* (Central City, CO: D. G. Collier, 1872), 19–20 for the Superintendent's advice on school architecture.
2. Allen Nossaman, *Many More Mountains*, vol. 1: *Silverton's Roots* (Denver, CO: Sundance Publications, 1989), 276, 271, and 254. However, a boy had been born to William H. and Carrie E. Nichols on October 18, 1874 in what was then La Plata County before Silverton had been established.
3. Ibid., 254; Helen M. Searcy, "Otto Mears," in Sarah Platt Decker Chapter D.A.R., *Pioneers of the San Juan Country*

(Colorado Springs, CO: Out West Printing and Stationary Co., 1942), 1:30.

4. According to Nossaman, vol. 2: *Ruts into Silverton* (Denver, CO: Sundance Publications, 1993), 27, both Mrs. Huff and her successor, Will Newcomb, had fifteen pupils in their classroom. The official census and enrollment figures as given in the State Superintendents reports cite eighteen boys and fifteen girls for 1875. Of these, twelve each supposedly attended Mrs. Huff's class. In 1876, the census listed twenty boys and eighteen girls of whom twelve boys and eleven girls were listed as attending Mr. Newcomb's instruction. In all likelihood, Mr. Nossaman's figures are closer to reality than those of the Superintendent.

5. Nossaman, 1.255 and 303.

6. Nossaman, 2:27.

7. Nossaman, 1:255, 276, 302, 2.27.

8. Nossaman, 2:12, 17, 25, 27, and 30.

9. Ibid., 28, 35, 61, and 115. For more on this subject see Anne Seagraves, *Soiled Doves: Prostitution in the Early West* (Hayden, ID: Wesanne Publications, 1994).

10. Nossaman, 2:115, 279, 12, and 38.

11. Nossaman, 1:290–291, 302, 331; 2:17; Charles Pinkerton, "The Pinkertons of Pinkerton Springs," in Sarah Platt Decker Chapter D.A.R., *Pioneers of the San Juan Country*, (Colorado Springs, CO: Out West Printing and Stationary Co., 1942), 2:79; Freda Carley Peterson, *The Story of Hillside Cemetery, Silverton, San Juan County, Colorado* (Silverton, CO: F.C. Peterson), 1:C-50–51.

12. Nossaman, 1:322.

13. Nossaman, 2:60, 65, 66, 83.

14. Nossaman, 1:223, 235–236; 2:21, 27, 62–63, 142. On Miss Puckett see *The La Plata Miner*, December 30, 1882 and *The San Juan Prospector*, June 16, 1877.

15. Nossaman, 2:142, 271.

16. Ibid., 142–143, 233. On Helen Standish Bowman see Nossaman, 1:313, and on Emma Hollingsworth see Peterson, *The Story of Hillside Cemetery*, 1:H-45.

17. Territory of Colorado, *Third Biennial Report of the Superintendent of Public Instruction for the Two YearsEending September 30th, 1875* (Denver, CO: Rocky Mountain News Steam Printing House, 1876), 98.

18. On this see my "Nineteenth Century Schools Between Community and State: The Cases of Prussia and the United States," *History of Education Quarterly*, 42 (Fall 2002), 317–341.

19. State of Colorado, *Fifth Biennial Report of the Superintendent of Public Instruction for the Years Ending August 31, 1885 and August 31, 1886* (Denver, CO: Collier and Cleaveland Lith. Co., State Printers, 1887).

20. State of Colorado, *First Biennial Report*, 9, 29.

21. Ibid., 11.

3   Town and School in a Wilderness

1. Allen Nossaman, *Many More Mountains, Vol. I: Silverton's Roots* (Denver, CO: Sundance Publications, 1989), 254; Helen M. Searcy, "Otto Mears," in Sarah Platt Decker Chapter D.A.R., *Pioneers of the San Juan Country* (Colorado Springs, CO: Out West Printing and Stationary Co., 1942), vol. 1, 30.

2. Allen Nossaman, *Many More Mountains*, vol. 2: *Ruts into Silverton* (Denver, CO: Sundance Publications, 1993), 86.

3. Allen Nossaman, *Many More Mountains*, vol. 3: *Rails into Silverton* (Denver, CO: Sundance Publications, 1998), 280, 295; W.H. Wigglesworth, "The Pathfinder of the San Juan," in *Pioneers of the San Juan Country*, 3:1–3.

4. Louisa Weinig Morgan, "Weinigs of the Vienna Restaurant," in Sarah Platt Decker Chapter D.A.R., *Pioneers of the San Juan Country*, (Colorado Springs, CO: Out West Printing and Stationary Co., 1942), 2:132.

5. Peter Scott, "Mr. Peter Scott's Story," in Platt Decker Chapter D.A.R., *Pioneers of the San Juan Country*, 1:47.

6. Nossaman, 2:132, 239.

7. Ibid., 242, 243.

8. Ibid., 280, 282.

9. See Michael Kaplan, *Otto Mears: Paradoxical Pathfinder* (Silverton, CO: San Juan County Book Co., 1982), 40.

10. *La Plata Miner*, October 11, 1879; Freda Ambold Lane, "The Gus A. Ambold Family," in Platt Decker Chapter D.A.R., *Pioneers of the San Juan Country*, 2:110. See also Jonathan Thompson's vivid description of what came to be known as the Lime Creek Burn of 1879 in "A Scene of Weird Magnificence," in *Journey through Time*, a special publication of the *Silverton Standard* and the *Miner* (Summer 2004).

11. Nossaman, 2:166.
12. Nossaman, 3:65–66; Allan G. Bird, *Bordellos of Blair Street: The Story of Silverton, Colorado's Notorious Red Light District*, rev. ed. (Pierson, MI: Advertising, Publications & Consultants, 1993), 4–9. Nossaman and Bird ascribe the passing of the prostitution ordinance to different years. According to Nossaman, it was passed in 1881. Bird cites 1879.
13. Nossaman, 2:162.
14. Ibid., 233, 238.
15. Morgan, 2:132.
16. *La Plata Miner*, May 5, 1879.
17. Andrew Gulliford, *America's Country Schools* (Washington, DC: Preservation Press, 1984), 70; State of Colorado, *First Biennial Report of the Superintendent of Public Instruction for the Two Years Ending August 31,1878* (Denver, CO: Daily Times Printing House and Book Manufactury, 1879), 15.
18. Nossaman, 2:271.
19. *La Plata Miner*, September 27, 1879; Jurgen Herbst, *And Sadly Teach* (Madison, WI: University of Wisconsin Press, 1989), 118; Nossaman, 2:271.
20. *La Plata Miner*, October 18, 1879; Nossaman, 2:288, 305.
21. *La Plata Miner*, September 13 and October 4, 1879.
22. *La Plata Miner*, July 5 and September 20, 1879.
23. *La Plata Miner*, November 1 and 8, 1879.
24. Nossaman, 2:290.

4 A SETTLEMENT TAKES HOLD

1. Quoted in John Marshall with Zeke Zanoni, *Mining the Hard Rock in the Silverton San Juans: A Sense of Place a Sense of Time* (Silverton, CO: Simpler Way Book Co., 1996), 16.
2. Allen Nossaman, *Many More Mountains, Vol. III: Rails into Silverton* (Denver, CO: Sundance Publications, 1998), 261, 263, 301.
3. The population statistics given in this chapter are taken from the 1880 Federal Census of San Juan County. See table 4.1
4. *La Plata Miner*, January 1, 1881.
5. Allen Nossaman, *Many More Mountains, Vol. II: Ruts into Silverton* (Denver, CO: Sundance Publications, 1993), 336.
6. Marshall and Zanoni, *Mining the Hard Rock*, 6–7.
7. *La Plata Miner*, December 25, 1880.

8. *San Juan Herald*, January 26, February 23, and March 16, 1882.
9. Nossaman, 2:334; 3:34, and *La Plata Miner*, August 20, 1880 and July 16, 1881.
10. Nossaman, 3:191.
11. Ibid., 34–38.
12. Nossaman, 2:296.
13. Helen M. Searcy, "Otto Mears," in Sarah Platt Decker Chapter D.A.R., *Pioneers of the San Juan Country*, (Colorado Springs, CO: Out West Printing and Stationary Co., 1942), 1:31–32, and Michael Kaplan, *Otto Mears: Paradoxical Pathfinder* (Silverton, CO: San Juan County Book Co., 1982), 63–67.
14. Nossaman, 3:174, 184.
15. Ibid., 214, 233.
16. For more on Silverton's bordellos see Nossaman, 2:314 and Allan G. Bird, *Bordellos of Blair Street: The Story of Silverton, Colorado's Notorious Red Light District*, rev. ed. (Pierson, MI: Advertising, Publications and Consultants, 1993), 6–9.
17. Elliott West, *Growing Up with the Country: Childhood on the Far Western Frontier* (Albuquerque, NM: University of New Mexico Press, 1989), 127–128.
18. On the basis of accounts given in Bird, *Bordellos of Blair Street*, 11–16; *La Plata Miner* (December 30, 1882); and Annie R. Gray, "Experiences in the San Juan Country," in Platt Decker Chapter D.R.A., *Pioneers of the San Juan Country*, 4:125. A more extensive description of the "Midnight Murder" may be found in *Journey through Time*, a special publication of the *Silverton Standard and the Miner* (Summer 2004).
19. *La Plata Miner*, March 5, 1881.
20. Minutes, Board of Directors, School District No. 1, June 4, 1881.
21. Minutes, Board of Directors, September 1, 1881.
22. The school census was reported in the *La Plata Miner* on July 23 and the October school report appeared in the *San Juan Herald* of October 22, 1881.
23. *La Plata Miner*, June 24, 1882.
24. *La Plata Miner*, May 1, 1880.
25. *La Plata Miner*, May 8, 1880.
26. State of Colorado, *Second Biennial Report of the Superintendent of Public Instruction for the Years Ending August 31, 1879 and August 31, 1880* (Denver, CO: Tribune Publishing Co., State Printers, 1881), 23.

27. *La Plata Miner*, November 6 and 13, 1880.
28. "Stand by the Schools," *La Plata Miner*, March 5, 1881.
29. Minutes, Board of Directors, May 18 and June 4, 1881.
30. Minutes, Board of Directors, June 27, 1881.
31. *The San Juan Herald*, September 15 and November 10, 1881.
32. *La Plata Miner*, November 12, 1881.
33. *The San Juan Herald*, April 20, 1882.

5 At High Tide

1. In August 1882 the San Juan County Bank and the Bank of Silverton merged.
2. *La Plata Miner*, June 23, 1883.
3. See *San Juan Herald* of October 22, 1881, and *La Plata Miner* of September 22, 1883.
4. *The San Juan Herald*, September 1, 1881 and November 24, 1881.
5. Allen Nossaman, *Many More Mountains, Vol. III: Rails into Silverton* (Denver, CO: Sundance Publications, 1998), 114–115.
6. State of Colorado, *Fifth Biennial Report of the Superintendent of Public Instruction for the Years Ending August 31, 1885 and August 31, 1886* (Denver, CO: Collier and Cleaveland Lith. Co., State Printers, 1887), 75.
7. As cited in the *San Juan Herald* of January 12, 1882.
8. *La Plata Miner*, May 13 and June 9, 1882; Nossaman, 3:259.
9. *La Plata Miner*, September 1, 1883.
10. *The Silverton Democrat*, May 3 and 17, 1884; "The School House Question," *La Plata Miner*, May 24, 1884.
11. *La Plata Miner*, November 22, 1884 and January 10, 1885.
12. *The Silverton Democrat*, December 13, 1884.
13. *The Silverton Democrat*, March 7, 1885. A check in a 1973 edition of Webster's New Collegiate Dictionary found no such entry.
14. *The Silverton Democrat*, May 16, 1885.
15. *La Plata Miner*, November 10, 1883.
16. *The Silverton Democrat*, September 20, 1884.
17. *The San Juan Herald*, December 18, 1884.
18. *La Plata Miner*, July 18 and August 28, 1885.
19. *La Plata Miner*, March 17, June 30, and October 20, 1883.
20. *The San Juan Herald*, July 26 and September 6, 1883.
21. *La Plata Miner*, June 23 and July 7, 1883.

22. *La Plata Miner,* July 7, 1883.
23. *La Plata Miner,* September 22, 1883.
24. *The San Juan Herald,* September 6, 1883.
25. *La Plata Miner,* June 23, 1883.
26. *La Plata Miner,* July 7 and November 24, 1883.
27. *The San Juan Herald,* May 25 and July 6, 1882, and October 11, 1883.
28. *La Plata Miner,* February 23, 1884.
29. Helen M. Searcy, "The Big Snow of 1884," in Sarah Platt Decker Chapter D.A.R., *Pioneers of the San Juan Country* (Colorado Springs, CO: Out West Printing and Stationary Co., 1946) 2:182–183; *La Plata Miner,* February 9 and 16, 1884.
30. *La Plata Miner,* April 5, 1884.
31. *La Plata Miner,* February 16, 1884.
32. *The Silverton Democrat,* September 5, 1885.
33. *La Plata Miner,* March 22, 1884.
34. *The San Juan Herald,* April 17, 1884.

6 School and Town in the 1880s

1. *La Plata Miner,* June 26, 1886.
2. By 1887 the Bank of Silverton had been declared insolvent and in 1888 the First National Bank was the only Bank in Silverton. The Bank of Silverton was to reopen in 1892 only to fail again in 1903.
3. *The San Juan,* January 8, 1888.
4. *La Plata Miner,* February 13, 1886.
5. *La Plata Miner,* February 20, 27, and June 6, 1886; *The Silverton Democrat,* September 21, 1886.
6. *La Plata Miner,* February 20, 1886.
7. *La Plata Miner,* May 1 and 22, 1886.
8. *La Plata Miner,* March 13, 1886.
9. *La Plata Miner,* March 27, 1886.
10. *The San Juan,* November 11, 1886, and April 14, 1887.
11. *The San Juan,* November 25 and December 23, 1886; Helen M. Searcy, "Otto Mears" and Josie Moore Crum, "Auxiliary Railroads in the San Juan Basin," in Sarah Platt Decker Chapter D.A.R., *Pioneers of the San Juan Country* (Colorado Springs, CO: Out West Printing and Stationary Co., 1942), 1:34–35, 180.

12. *La Plata Miner*, April 10, 1886; *The San Juan*, November 18 and 25, 1886.
13. *The San Juan*, January 20, 1887.
14. *The San Juan*, March 31, May 19, 26, June 9, and 30, 1887.
15. *The San Juan*, August 25, 1887.
16. *The Silverton Democrat*, October 6, 1887.
17. Selwyn K. Troen, *The Public and the Schools: Shaping the St. Louis System, 1838–1920* (Columbia, MO: University of Missouri Press, 1975), 75–77; *The San Juan*, November 24, 1887.
18. *The San Juan*, January 12 and February 23, 1888.
19. *La Plata Miner*, June 12, 1886; Minutes of the School Board, August 4, 1886.
20. Minutes of the School Board, August 24 and December 22, 1886.
21. *The San Juan*, January 20, 1887; *The San Juan Democrat*, June 21, 1888; *The Silverton Democrat*, December 11, 1886.
22. *The San Juan*, February 3, March 3, 10, 31, and May 12, 1887; *The Silverton Democrat*, February 12 and March 12, 1887.
23. *The Silverton Democrat*, April 23 and June 2, 1887; Minutes of the School Board, April 26, 1887; *The San Juan*, May 12, 1887.
24. *The San Juan*, July 28 and August 18, 1887.
25. *The San Juan*, April 28 and May 12, 1887; *The Silverton Democrat*, May 7, 1887.
26. Minutes of the School Board, July 23, 1887; *The San Juan*, November 17, 1887.
27. *The San Juan*, January 5, 1888.

7 A School in Crisis

1. Minutes of the School Board, March 10, 1888.
2. *The San Juan Democrat*, May 10 and August 30, 1888; Minutes of the School Board, August 28, 1888.
3. *The San Juan Democrat*, August 30 and September 6, 1888.
4. *The San Juan Democrat*, September 13, 20, November 1 and 8, 1888; Minutes of the School Board, November 24, 1888.
5. Minutes of the School Board, December 18, 1888, May 6, 10, and July 10, 1889.
6. *The Silverton Standard*, November 2, 1889; April 12 and May 10, 1890.

7. Minutes of the School Board, February 13, 1889. See Silverton Public Schools, *Rules for Pupils and Rules for Teachers* (see appendix V and VI).
8. *The Silverton Standard*, January 4, 1890.
9. *The Silverton Weekly Miner*, June 28 and August 16, 1890; *The Silverton Standard*, July 12, 1890.
10. Minutes of the School Board, June 14, 1890; *The Silverton Standard*, June 21, 1890.
11. *The Silverton Standard*, January 31, 1891.
12. In State of Colorado, *Eighth Biennial Report of the Superintendent of Public Instruction for the Two Years Ending June 30, 1892* (Denver, CO: Smith-Brooks Printing Co., State Printers, 1893), 832.
13. State of Colorado, *Twelfth Biennial Report of the Superintendent of Public Instruction for the Two Years Ending November 15, 1900* (Denver, CO: Smith-Brooks Printing Co., State Printers, 1900), 15.
14. *The Silverton Weekly Miner*, April 18 and 25, 1891.
15. State of Colorado, *Seventh Biennial Report of the Superintendent of Public Instruction for the Biennial Term Ending June 30, 1890* (Denver, CO: Collier and Cleaveland Lith. Co., Printers, 1891), 88.
16. *The Silverton Standard*, April 18, 1891.
17. Freda Carley Peterson, *The Story of Hillside Cemetery, Silverton, San Juan County, Colorado* (Silverton, CO: F.C. Peterson, 1996), 1:H-45.
18. *The Silverton Weekly Miner*, April 25 and May 2, 1891.
19. *The Silverton Standard*, May 2, 1891.
20. *The Silverton Standard*, June 20 and 27, 1891; *The Silverton Weekly Miner*, June 27, 1891; Minutes of the School Board, July 1, 1891.
21. *The Silverton Weekly Miner* and *The Silverton Standard*, July 25, 1891.
22. Minutes of the School Board, March 14 and April 1, 1892; *The Silverton Standard*, April 2, 1892.
23. *The Silverton Standard*, April 9, 16, and May 7, 1892. It might have been illuminating to see the *Miner*'s account of the events, but unfortunately copies for this period appear to be missing.
24. Minutes of the School Board, July 6 and 19, 1892; *The Silverton Standard*, September 3, 1892.

25. Minutes of the School Board, August 29, 1892, and *The Silverton Standard*, September 24, 1892.
26. *The Silverton Standard*, October 1, 1892.
27. Minutes of the School Board, October 3 and October 10, 1892, December 26, 1894, and February 4, 1895.

8 THE TURBULENT 1890S

1. Samuel Eliot Morison and Henry Steele Commager, *The Growth of the American Republic* (New York: Oxford University Press, 1962), 2:334; *Documents of American History*, ed. Henry Steele Commager, 7th ed. (New York: Appleton-Century-Crofts, 1963), 547, 587.
2. Cleveland's Silver Letter, February 10, 1892 in Commager, *Documents*, 588.
3. *The Silverton Weekly Miner*, July 5, 1890.
4. *The Silverton Weekly Miner*, August 30 and December 27, 1890.
5. *The Silverton Standard*, February 26, May 7, 1892, and January 23, 1893.
6. *The Silverton Standard*, March 14, 1891.
7. *The Silverton Weekly Miner*, April 8, 1890; *The Silverton Standard*, August 9, 1890 and April 25, 1891.
8. *The Silverton Standard*, March 5, 1892; Allan G. Bird, *Bordellos of Blair Street: The Story of Silverton, Colorado's Notorious Red Light District*, rev. ed. (Pierson, MI: Advertising, Publications and Consultants, 1993), 7–8, 94–95, 131–132.
9. *The Silverton Standard*, March 12, 1892.
10. *The Silverton Standard*, July 8 and 15, 1893; Duane A. Smith, *Song of the Hammer and Drill: The Colorado San Juans, 1860–1914* (Boulder, CO: University Press of Colorado, 2000), 129; Commager, *Documents*, 599.
11. *The Silverton Standard*, July 22, August 5, and 26, 1893.
12. *The Silverton Weekly Miner*, March 2, 1894; *The Silverton Standard*, April 21, 1894 and March 2, 1895.
13. *The Silverton Weekly Miner*, February 9, 1894, February 1, and March 1, 1895.
14. Smith, *Song of the Hammer and Drill*, 127–130.
15. *The Silverton Standard*, May 25 and November 9, 1895; *The Silverton Weekly Miner*, November 15, 1895.
16. Reprinted in the *Silverton Standard*, December 28, 1895.

17. *The Silverton Weekly Miner*, April 8, 1890; *The Silverton Standard*, September 27 and October 4, 1890.

18. *The Silverton Weekly Miner*, October 18 and 25, November 1, 1890; *The Silverton Standard*, November 8, 1890.

19. *The Silverton Standard*, April 4 and 11, August 1, 8, and 15, 1891, February 26, 1892; *The Silverton Weekly Miner*, April 18, 1891.

20. *The Silverton Standard*, August 26, October 14, and November 11, 1893; quoted in the *Silverton Standard*, December 9, 1893.

21. *The Silverton Standard*, March 24 and April 7, 1894; *The Silverton Weekly Miner*, April 6, 1894.

22. *The Silverton Weekly Miner*, March 2, 1894 and November 19, 1897.

23. *The Silverton Standard*, April 6 and October 19, 1895.

24. Josie Moore Crum, "Auxiliary Railroads in the San Juan Basin," in Sarah Platt Decker Chapter D.A.R., *Pioneers of the San Juan Country* (Colorado Springs, CO: Out West Printing and Stationary Co., 1942), 1:180–181; *The Silverton Weekly Miner*, January 8, 1897, July 29, 1898, May 19, and November 3, 1899.

25. *The Silverton Standard*, March 20, 1897; *The Silverton Weekly Miner*, June 4, 1897 and July 22, 1898.

26. I have borrowed here heavily from Smith, *Song of the Hammer and Drill*, 138–141.

27. *The Silverton Weekly Miner*, September 25, 1896, July 16, and August 27, 1897.

28. *The Silverton Standard*, September 4, 1897; *The Silverton Weekly Miner*, October 28, 1898.

29. *The Silverton Weekly Miner*, April 30 and May 21, 1897; *The Silverton Standard*, February 2, 1895, February 20, 27, and August 14, 1897.

30. *The Silverton Weekly Miner*, June 8 and 15, 1894, October 20, 1899, and November 9, 1900.

31. *The Silverton Standard*, March 4, June 10, and July 1, 1899.

32. *The Silverton Standard*, July 22 and August 5, 1899; *The Silverton Weekly Miner*, August 11, 1899.

33. *The Silverton Standard*, August 19, 1899; *The Silverton Weekly Miner*, September 15 and 22, 1899.

9   SILVERTON'S LIFE AT CENTURY'S END

1. *The Silverton Weekly Miner*, March 2, 1894; *The Silverton Standard*, June 8, 1895.

2. *The Silverton Weekly Miner,* September 7, 1894 and April 5, 1895; Minutes of the School Board, November 14, 1894 and March 4, 1895.

3. *The Silverton Weekly Miner,* June 21, 1895.

4. See my *The Once and Future School: Three Hundred and Fifty Years of American Secondary Education* (New York: Routledge, 1996), 53–54.

5. Minutes of the School Board, June 24, 1895; *The Silverton Weekly Miner,* June 28 and August 16, 1895.

6. The American Protective Association was an anti-Catholic and anti-immigrant Association.

7. *The Silverton Standard,* September 7, 1895 and January 18, 1896; Minutes of the School Board, September 2, 1895; *The Silverton Weekly Miner,* August 9, September 13 and 29, 1895.

8. *The Silverton Standard,* December 5, 1896 and February 6, 1897.

9. *The Silverton Weekly Miner,* April 30, 1897; *The Silverton Standard,* May 1 and 8, 1897.

10. *The Silverton Weekly Miner,* May 28, June 4, and 18, 1897.

11. Minutes of the School Board, August 28 and September 6, 1897.

12. *The Silverton Standard,* June 4 and July 30, 1898; Minutes of the School Board, September 5, 1898.

13. Minutes of the School Board, May 6 and 26, June 24, July 3, August 7 and 28, and September 4, 1899.

14. *The Silverton Standard,* October 21, November 25, 1899, January 20, 1900; *The Silverton Weekly Miner,* September 29 and October 20, 1899, and May 5, 1900.

15. Report Card and Memories of Life in Silverton by courtesy of San Juan County Historical Society.

16. *The Silverton Weekly Miner,* November 22 and December 27, 1895.

17. *The Silverton Weekly Miner,* December 27, 1895, January 24 and May 15, 1896, and August 6, 1897.

18. Quoted in Samuel Eliot Morison and Henry Steele Commager, *The Growth of the American Republic* (New York: Oxford University Press, 1962), 2:357.

19. *The Silverton Weekly Miner,* September 25, 1896; *The Silverton Standard,* October 31, 1896.

20. *The Silverton Weekly Miner,* March 12, 27, April 2 and 9, 1897; *The Silverton Standard,* April 17, May 8, and September 18, 1897.

21. *The Silverton Weekly Miner*, October 22, 1897; *The Silverton Standard*, October 23, 1897.
22. *The Silverton Weekly Miner*, April 1 and 8, 1898.
23. *The Silverton Weekly Miner*, September 6 and October 14, 1898; *The Silverton Standard*, October 22 and 29, 1898.
24. *The Silverton Weekly Miner*, March 10 and April 7, 1899; *The Silverton Standard*, March 18 and April 22, 1899.
25. *The Silverton Standard*, September 2 and October 28, 1899; *The Silverton Weekly Miner*, October 20 and November 10, 1899.
26. *The Silverton Standard*, January 26, February 17, 24, March 10, and April 7, 1900; *The Silverton Weekly Miner*, March 23, 30, and April 6, 1900.
27. Freda Carley Peterson, *The Story of Hillside Cemetery, Silverton, San Juan County, Colorado* (Silverton, CO: F.C. Peterson, 1996) 1:L 2.
28. Morison and Commager, *The Growth of the American Republic*, 2:422–427.
29. *The Silverton Standard*, April 16 and 23, 1898.
30. *The Silverton Standard*, August 6, 1898.
31. *The Silverton Standard*, December 24, 1898 and December 30, 1899.
32. *The Silverton Standard*, April 29 and May 13, 1899.
33. *The Silverton Weekly Miner*, August 11, 1899.

10   SILVERTON ENTERS THE TWENTIETH CENTURY

1. See the *Silverton Standard* of January 3, 1903, January 2, 1904, and December 31, 1904.
2. Duane Smith, *Song of the Hammer and Drill: The Colorado San Juans, 1860–1914* (Boulder, CO: University Press of Colorado, 1982), 130–135, 182; *The Silverton Standard*, August 24, 1901.
3. Smith, *Song of the Hammer*, 178, 179; *The Silverton Weekly Miner*, August 28, 1903; *The Silverton Standard*, September 5, 1903, and March 19, 1904.
4. *The Silverton Standard*, October 18, 1902 and September 5, 1903.
5. *The Silverton Weekly Miner*, March 6 and September 25, 1903.
6. Reprinted in the *Silverton Weekly Miner*, July 8, 1904.

7. *The Silverton Standard,* June 16, July 28, September 22 and 29, and November 17, 1900.

8. *The Silverton Weekly Miner,* August 24 and December 7, 1900 and September 13, 1901.

9. *The Silverton Standard,* May 17, 1902; *The Silverton Weekly Miner,* January 16, 26 and 30, 1903.

10. Freda Carley Peterson, *The Story of Hillside Cemetery, Silverton, San Juan County, Colorado* (Silverton, CO: F.C. Peterson, 1998), 2:R-24–26, and August Fast, "What I Remember," in Sarah Platt Decker Chapter D.A.R., *Pioneers of the San Juan Country* (Colorado Springs, CO: Out West Printing and Stationary Co., 1946), 2:171.

11. *The Silverton Standard,* February 17 and April 7, 1900.

12. *The Silverton Standard,* August 17, 1901; *The Silverton Weekly Miner,* September 6, 1901; *The Silverton Standard,* September 21, 1901.

13. *The Silverton Standard,* September 15, 1900; *The Silverton Weekly Miner,* March 29, 1901; *The Silverton Standard,* March 30 and April 6, 1901.

14. *The Silverton Weekly Miner,* March 21 and 28, 1902; *The Silverton Standard,* April 5, 1902, January 3 and October 10, 1903, January 2, 1904. See also Duane Smith, *Silverton: A Quick History* (Fort Collins, CO: First Light Publishing, 1997), 68.

15. *The Silverton Weekly Miner,* July 6, 1900; *The Silverton Standard,* August 11, 1900.

16. *The Silverton Standard,* November 10, 1900.

17. *The Silverton Standard,* April 5, 1902 and March 28, 1903; *The Silverton Weekly Miner,* April 5, 1902 and April 3, 1903.

18. *The Silverton Weekly Miner,* April 7, and *The Silverton Standard,* April 8, 1905.

19. *The Silverton Weekly Miner,* November 7, 1902 and March 24, 1905; *The Silverton Standard,* November 7, 1903 and November 12, 1904.

20. *The Silverton Standard,* December 29, 1900.

21. *The Silverton Standard,* February 2 and April 20, 1901.

22. *The Silverton Standard,* September 7, 1901, June 21, 1902, May 30, 1903, and January 2, 1904.

23. *The Silverton Weekly Miner,* October 4, 1901.

24. *The Silverton Weekly Miner,* February 7, 1902.

25. *The Silverton Standard,* February 8, 15, and 22, 1902.

26. *The Silverton Weekly Miner*, May 16, 1902; *The Silverton Standard*, May 17, 1902.
27. *The Silverton Standard*, May 17, 1902.
28. *The Silverton Weekly Miner*, May 23, 1902.
29. *The Silverton Weekly Miner*, February 17, 1905.

11   PUBLIC SCHOOLING'S TRIUMPH: HIGH SCHOOL GRADUATION

1. Statistics have been assembled from school reports printed in Silverton newspapers and, most importantly, from the school census files preserved in Silverton's public library.
2. See State of Colorado, *Twelfth Biennial Report of the Superintendent of Public Instruction of the State of Colorado for the Two Years Ending November 15, 1900* (Denver, CO: Smith-Brooks Printing Co., State Printers, 1900), 15.
3. *The Silverton Standard*, April 21 and May 5, 1900.
4. *The Silverton Standard*, May 7 and June 23, 1900.
5. Minutes of the School Board, September 3, 1900; *The Silverton Standard*, October 6 and 27, 1900.
6. *The Silverton Standard*, November 17 and December 1, 1900.
7. *The Silverton Weekly Miner*, December 7, 1900.
8. Minutes of the School Board, January 30 and February 6, 1901.
9. *The Silverton Standard*, February 9 and 23, 1901.
10. Minutes of the School Board, May 21 and 28, June 14, July 8 and 10, and September 9, 1901; *The Silverton Standard*, July 13 and August 3 and 31, 1901; *The Silverton Weekly Miner*, August 30, and September 13, 1901.
11. *The Silverton Weekly Miner*, November 1, 8, and 29, 1901.
12. *The Silverton Weekly Miner*, December 13, and *The Silverton Standard*, December 14, 1901 and April 5, 1902.
13. *The Silverton Weekly Miner*, February 21 and 28, 1902; *The Silverton Standard*, March 22, 1902.
14. *The Silverton Weekly Miner*, February 28 and March 14, 1902, and June 30, 1905.
15. *The Silverton Standard*, March 1 and May 3, 1902; *The Silverton Weekly Miner*, May 2 and 30, 1902.
16. For more on the American high school around 1900 see Theodore R. Sizer, *Secondary Schools at the Turn of the Century*

(New Haven, CT: Yale University Press, 1964) and my *The Once and Future School: Three Hundred and Fifty Years of American Secondary Education* (New York: Routledge, 1996).

17. See on this Freda Carley Peterson, *The Story of Hillside Cemetery, Silverton, San Juan County, Colorado* (Silverton, CO: F.C. Peterson, 1998), 2:C-8.

POSTSCRIPT: LOOKING BACKWARD AND LOOKING FORWARD

1. *The San Juan Herald*, December 18, 1884.
2. Elliott West, *Growing Up with the Country: Childhood on the Far Western Frontier* (Albuquerque, NM: University of New Mexico Press, 1989), 127–128.
3. *The Silverton Standard*, February 15, 1902.
4. Duane Smith, *Silverton: A Quick History* (Fort Colline, CO: First Lightning Publishing, 1997), 75.
5. I have relied here heavily on Smith, *Silverton*, 95, 104, 109.
6. George L. Mosse, *Confronting History: A Memoir* (Madison, WI: University of Wisconsin Press, 2000), 55.
7. Ibid., 58
8. Brian Simon, *A Life in Education* (London: Lawrence & Wishart, 1998), 8 and 9.
9. Karl Schwarz, *Die Kurzschulen Kurt Hahns: Ihre pädagogische Theorie und Praxis* (Ratingen bei Düsseldorf, Germany: A. Henn Verlag, 1968), 51–52.
10. Henry Brereton, "Kurt Hahn of Gordonstoun," in Hermann Röhrs and Hilary Tunstall-Behrens, eds., *Kurt Hahn* (London: Routledge & Kegan Paul, 1970), 57.
11. Schwarz, *Die Kurzschulen*, 52.
12. Ibid., 55; James M. Hogan, "The Establishment of the First Outward Bound School at Aberdovey, Merionethshire," in Röhrs and Tunstall-Behrens, *Kurt Hahn*, 65.
13. Josh Miner and Joe Boldt, *Outward Bound, USA: Crew Not Passengers* (Seattle, WA: Mountaineers Books, 2002), 84–88.
14. Joshua Miner, "Outward Bound in the USA," in Röhrs and Tunstall-Behrens, *Kurt Hahn*, 200–203.
15. Miner and Boldt, *Outward Bound*, 369.
16. The design principles may be found in Emily Cousins, ed., *Reflections on Design Principles* (Dubuque, IA: Kendall/Hunt

Publishing Co., 1998); Eleanor Duckworth, *The Having of Wonderful Ideas and Other Essays on Teaching and Learning* (New York: Teachers College Press, 1996).

17. Quoted in Miner and Boldt, *Outward Bound*, 373.
18. On Hahn's "experience therapy" see Schwarz, *Die Kurzschulen*, 41–44 and Röhrs, "The Educational Thought of Kurt Hahn," in Röhrs and Tunstall-Behrens, *Kurt Hahn*, 126–127.
19. Silverton School District, *2004–2005 Report to the Public* (Winter 2006), 4.
20. Silverton School District, *Annual Report to the Public, 2005–2006 School Year in Review* (2007), 2; and Ibid., *2004–2005 Report*, 1.
21. Ibid., Annual Report, 2005–2006, 2.

BIBLIOGRAPHY

BOOKS

Backus, Harriet Fish. *Tomboy Bride*. Boulder, CO: Pruett, 1969.
Benham, Jack Luther. *Silverton and Neighboring Ghost Towns*. Ouray, CO: Bear Creek, 1977.
Bird, Allan G. *Bordellos of Blair Street: The Story of Silverton, Colorado's Notorious Red Light District*. Rev. ed. Pierson, MI: Advertising, Publications and Consultants, 1993.
Celis, William. *Battle Rock: The Struggle over a One-Room School in America's Vanishing West*. New York: Public Affairs, 2002.
Colorado Families: A Territorial Heritage. Denver, CO: Colorado Genealogical Society, 1981.
Colville, Ruth Marie. *Del Norte, Colorado: Gateway to the San Juan*. Monte Vista, CO: San Luis Valley Publishing Co., 1987.
Cousins, Emily. *Reflections on Design Principles*. Dubuque, IO: Kendall/Hunt, 1998.
——— (ed.). *Roots: From Outward Bound to Expeditionary Learning*. Dubuque, IO: Kendall/Hunt, 2000.
Cousins, Emily and Amy Mednic. *Fieldwork: An Expeditionary Learning Outward Bound Reader*, vol. 2. Dubuque, IO: Kendall/Hunt, 1996.
Cousins, Emily and Melissa Rodgers. *Fieldwork: An Expeditionary Learning Outward Bound Reader*, vol. 1. Dubuque, IO: Kendall/Hunt, 1995.
Duckworth, Eleanor. *The Having of Wonderful Ideas and Other Essays on Teaching and Learning*. New York: Teachers College Press, 1996.
Emmitt, Robert. *The Last War Trail: The Utes and the Settlement of Colorado*. Norman, OK: University of Oklahoma Press, 1954.
Etulain, Richard W. with Pat Devejian, Jon Hunner, and Jacqueline Etulain Partch (eds.). *The American West in the Twentieth Century*. Norman, OK: University of Oklahoma Press, 1994.

Fetchenbier, Scott. *Ghosts and Gold: The History of the Old One Hundred Mine*. Silverton, CO: Miner Commissary Gift Shop, 1999.

Flavin, Martin. *Kurt Hahn's School and Legacy*. Wilmington, DE: Middle Atlantic Press, 1996.

Foote, Mary Hallock. *A Victorian Gentlewoman in the Far West: The Reminiscences of Mary Hallock Foote*. Edited by Rodman W. Paul. San Marino, CA: Huntingdon Library, 1972.

Gulliford, Andrew. *America's Country Schools*. Washington, DC: Preservation Press, 1984.

Hall, Frank. *History of the State of Colorado*. 4 vols. Chicago, IL: Blakely Printing Co., 1889.

Herbst, Jurgen. *The Once and Future School: Three Hundred and Fifty Years of American Secondary Education*. New York: Routledge, 1996.

Historical Encyclopedia of Colorado. Denver, CO: Colorado Historical Association, 1968 reprint.

Horgan, Paul. *Great River: The Rio Grande in North American History*. New York and Toronto, Canada: Rinehart & Co., 1954.

James, Thomas. *Education at the Edge: The Colorado Outward Bound School*. Denver, CO: Colorado Outward Bound School, 1980.

Jefferson, James, Robert W. Delaney, and Gregory C. Thompson. *The Southern Utes: A Tribal History*. Ignacio, CO: Southern Ute Tribe, 1972.

Kaplan, Michael. *Otto Mears: Paradoxical Pathfinder*. Silverton, CO: San Juan County Book Co., 1982.

Marshall, John and Jerry Roberts. *Living (and Dying) in Avalanche Country: Stories from the San Juans of Southwestern Colorado*. Silverton, CO: J. Marshall, 1992.

Marshall, John with Zeke Zanoni. *Mining the Hard Rock in the Silverton San Juans: A Sense of Place a Sense of Time*. Silverton, CO: Simpler Way Book Co., 1996.

Matheson, Samuel James. "A History of Public Schools in Colorado: 1859–1880." Doctoral thesis, University of Denver, 1963.

Miner, Joshua L. and Joe Boldt. *Outward Bound, U.S.A.: Crew Not Passengers*. 2nd ed. Seattle, WA: Mountaineers Books, 2002.

Morison, Samuel Eliot, and Henry Steele Commager. *The Growth of the American Republic*. 2 vols. New York: Oxford University Press, 1962.

Mosse, George L. *Confronting History: A Memoir*. Madison, WI: University of Wisconsin Press, 2000.

Nossaman, Allen. *Many More Mountains*, vol. 1: *Silverton's Roots*, vol. 2: *Ruts into Silverton*, vol. 3: *Rails into Silverton*. Denver, CO: Sundance, 1989–1998.

——— (comp.). *San Juan County Newspaper Index, 1884–1887*. Silverton, CO: Silverton Public Library and San Juan County Historical Society, 1985.

Oelkers, Jürgen. *Reformpädagogik: Eine kritische Dogmengeschichte*. Weinheim und München, Germany: Juventa, 1989.

Peterson, Freda Carley. *The Story of Hillside Cemetery, Silverton, San Juan County, Colorado*. 2 vols. Silverton, CO: F.C. Peterson, 1996–1998.

Pielorz, Anja. *Werte und Wege der Erlebnispädagogik Schule Schloss Salem*. Hamburg, Germany: Hermann Luchterhand Verlag, 1991.

Rathmell, Ruth. *Of Record and Reminiscence—Ouray and Silverton*. Ouray, CO: Ruth Rathmell, 1976.

Richards, Anthony. *Kurt Hahn: The Midwife of Educational Ideas*. Dr. Ed. Thesis, University of Colorado, 1981.

Rockwell, Wilson. *The Utes: A Forgotten People*. Denver, CO: Sage, 1956.

Röhrs, Hermann, and Hilary Tunstall-Behrens (eds.). *Kurt Hahn*. London: Routledge and Kegan Paul, 1970.

Sarah Platt Decker Chapter, D.A.R. *Pioneers of the San Juan Country*. 4 vols. Colorado Springs, CO: Out West Printing and Stationary Co., 1942–1961.

Schwartz, Karl. *Die Kurzschulen Kurt Hahns: Ihre pädagogische Theorie und Praxis*. Ratingen bei Düsseldorf, Germany: A. Henn Verlag, 1968.

Seagraves, Anne. *Soiled Doves: Prostitution in the Early West*. Hayden, ID: Wesanne, 1994.

Simon, Brian. *A Life in Education*. London: Lawrence and Wishart, 1998.

Sizer, Theodore R. *Secondary Schools at the Turn of the Century*. New Haven, CT: Yale University Press, 1964.

Smith, Duane. *Silverton: A Quick History*. Fort Collins, CO: First Light Publishing, 1997.

———. *Song of the Hammer and Drill: The Colorado San Juans, 1860–1914*. Boulder, CO: University Press of Colorado, 1982.

Stone, Wilbur. *History of Colorado*, vol. 4. Chicago, IL: S. J. Clarke Publishing Co., 1919.

Troen, Selwyn K. *The Public and the Schools: Shaping the St. Louis System, 1838–1920*. Columbia, MO: University of Missouri Press, 1975.

Ubbelohde, Carl. *A Colorado History*. Boulder, CO: Pruett, 1965.

West, Elliott. *Growing Up with the Country: Childhood on the Far Western Frontier*. Albuquerque, NM: University of New Mexico Press, 1989.

————. *The Saloon on the Rocky Mountain Mining Frontier*. Lincoln, NE: University of Nebraska Press, 1979.

West, Elliott and Paula Petrick (eds.). *Small Worlds: Children and Adolescents in America, 1850–1950*. Lawrence, KS: University Press of Kansas, 1992.

Wyman, Louis. *Snowflakes and Quartz: Stories of the Early Days in the San Juan Mountains*. Silverton, CO: Simpler Way Book Co., 1993.

ARTICLES

Aspaas, Hans. "The Aspaas Family," in Sarah Platt Decker Chapter, D.A.R., *Pioneers of the San Juan Country*, 1952), 3: 158–162.

Ayers, Mary C. "Howardsville in the San Juan," *Colorado Magazine*, 28, 4 (October 1951), 244–257.

Borland, Dr. Lois. "The Sale of the San Juan," *Colorado Magazine*, 28, 2 (April 1951), 107–127.

Brereton, Henry. "Kurt Hahn of Gordonstoun," in Röhrs and Tunstall-Behrens, , *Kurt Hahn*, 39–59.

Butchart, Ronald E. "Education and Culture in the Trans-Mississippi West: An Interpretation," *Journal of American Culture*, 3 (Summer 1980), 351–373.

————. "The Frontier Teacher: Arizona 1875–1925," *Journal of the West*, 16 (1977), 54–66.

Crum, Josie Moore. "Auxiliary Railroads in the San Juan Basin," in Sarah Platt Decker Chapter, D.A.R.., *Pioneers of the San Juan Country*, 1942, 1:180–181.

Espinosa, Fred. "Del Norte—Its Past and Present," *Colorado Magazine*, 5, 3 (June 1928), 95–102.

Fast, August. "What I Remember," inSarah Platt Decker Chapter, D.A.R., *Pioneers of the San Juan Country*, 1946, 2:153–172.

Gray, Annie R. "Experiences in the San Juan Country," in.Sarah Platt Decker Chapter,, D.A.R., *Pioneers of the San Juan Country*, 1961, 4:124–129.

"History of the San Juan County," *Colorado School of Mines Quarterly* (October 1910).

Hogan, James M. "The Establishment of the First Outward Bound School at Aberdovey, Merionetshire," in Röhrs and Tunstall-Behrens, *Kurt Hahn*, 60–66.

Knapp, Lena S. "John W. Wingate (Patriot)," in Sarah Platt Decker Chapter,, D.A.R., *Pioneers of the San Juan Country*, 1952, 3:148–152.

Lane, Freda Ambold. "The Gus A. Ambold Family," in Sarah Platt Decker Chapter, D.A.R., *Pioneers of the San Juan Country*, 1946, 2:110–111.

Miner, Joshua. "Outward Bound in the USA," in Röhrs and Tunstall-Behrens, *Kurt Hahn*, 197–208.

Morgan, Louisa Weinig. "Weinigs of the Vienna Restaurant," in Sarah Platt Decker Chapter, D.A.R., *Pioneers of the San Juan Country*, 1946, 2:132–136.

Pinkerton, Charles. "The Pinkertons of Pinkerton Springs," in Sarah Platt Decker Chapter, D.A.R. , *Pioneers of the San Juan Country*, 1946, 2:77–86.

Röhrs, H. "The Educational Thought of Kurt Hahn," in Röhrs and Tunstall-Behrens, *Kurt Hahn*, 123–136.

Ross, D. Reid. "Presbyterian Mission Schools and Teachers in the West After the Civil War," *Journal of the West*, 29 (October 1990), 82–93.

Schroeder, Albert H. "A Brief History of the Southern Utes," *Southwestern Lore*, 30, 4 (March 1965), 53–78.

Scott, Peter. "Mr. Peter Scott's Story," in Sarah Platt Decker Chapter, D.A.R., *Pioneers of the San Juan Country*, 1942, 1:47.

Searcy, Helen M. "Otto Mears," in Sarah Platt Decker Chapter, D.A.R., *Pioneers of the San Juan Country*, 1942, 1:15–47.

———— "The Big Snow of 1884," in Sarah Platt Decker Chapter, D.A.R., *Pioneers of the San Juan Country*, 1946, 2:182–189.

Taylor, John G. "When Judge Hallett Held the First Court in the San Juan." *Denver Times*, December 2, 1901.

Thompson, Jonathan. "A Scene of Weird Magnificence," *Journey through Time* (Silverton, CO: *The Silverton Standard* & *the Miner*, Summer 2004).

Wigglesworth, W.H. "The Pathfinder of the San Juan," in Sarah Platt Decker Chapter, D.A.R., *Pioneers of the San Juan Country*, 1952, vol. 3, 1–3.

GOVERNMENT PUBLICATIONS

Silverton School District. *2004–2005 Report to the Public*. Silverton, CO: 2005.

————. *Annual Report to the Public 2005–2006 School Year*. Silverton, CO: 2006.

State of Colorado, *First Biennial Report of the Superintendent of Public Instruction for the Two Years Ending August 31, 1878.* Denver, CO: Daily Times Printing House and Book Manufactury, 1879.

———. *Second Biennial Report of the Superintendent of Public Instruction for the Years Ending August 31, 1879 and August 31, 1880.* Denver, CO: Tribune, State Printers, 1881.

———. *Third Biennial Report of the Superintendent of Public Instruction for the Years Ending August 31, 1881 and August 31, 1882.* Denver, CO: Times Public Printer, 1883.

———. *Fourth Biennial Report of the Superintendent of Public Instruction for the Years Ending August 31, 1883 and August 31, 1884.* Denver, CO: Collier and Cleaveland, State Printers, 1885.

———. *Fifth Biennial Report of the Superintendent of Public Instruction for the Years Ending August 31, 1885 and August 31, 1886.* Denver, CO: Collier and Cleaveland, State Printers, 1887.

———. *Sixth Biennial Report of the Superintendent of Public Instruction for the Biennial Term Ending June 30, 1888.* Denver, CO: Collier and Cleaveland, State Printers, 1889.

———. *Seventh Biennial Report of the Superintendent of Public Instruction for the Biennial Term Ending June 30, 1890.* Denver, CO: Collier and Cleaveland , State Printers, 1891.

———. *Eighth Biennial Report of the Superintendent of Public Instruction for the Two Years Ending June 30, 1892.* Denver, CO: Smith-Brooks, State Printers, 1893.

———. *Ninth Biennial Report of the Superintendent of Public Instruction for the Two Years Ending June 30, 1894.* Denver, CO: Smith-Brooks, State Printers, 1895.

———. *Tenth Biennial Report of the Superintendent of Public Instruction for the Two Years Ending June 30, 1896.* Denver, CO: Smith-Brooks, State Printers, 1896.

———. *Eleventh Biennial Report of the Superintendent of Public Instruction for the Two Years Ending June 30, 1898.* Denver, CO: Smith-Brooks, State Printers, 1898.

———. *Twelfth Biennial Report of the Superintendent of Public Instruction for the Two Years Ending November 15, 1900.* Denver, CO: Smith-Brooks, State Printers, 1900.

———. *Thirteenth Biennial Report of the Superintendent of Public Instruction for the Two Years Ending November 30, 1902.* Denver, CO: Smith-Brooks, State Printers, 1902.

———. *Fourteenth Biennial Report of the Superintendent of Public Instruction for the Two Years Ending November 30, 1904.* Denver, CO: Smith-Brooks, State Printers, 1904.

Territory of Colorado. *First Biennial Report of the Superintendent of Public Instruction for the School Year Ending September 30, 1870 and September 30, 1871.* Central City, CO: D.G. Collier, 1872.

———. *Third Biennial Report of the Superintendent of Public Instruction for the Two Years Ending September 30th, 1875.* Denver, CO: Rocky Mountain News Steam Printing House, 1876.

PREVIOUSLY PUBLISHED BOOKS

The German Historical School in American Scholarship: A Study in the Transfer of Culture.
The History of American Education, Goldentree Bibliographies in American History.
From Crisis to Crisis: American College Government 1636–1819.
And Sadly Teach: Teacher Education and Professionalization in American Culture.
Research and Teaching: Personal Reflections and the University in the United States: Tradition and Reform.
The Once and Future School: Three Hundred and Fifty Years of American Secondary Education.
Requiem for a German Past: A Boyhood Among the Nazis. German edition: *Requiem für eine deutsche Vergangenheit.*
School Choice and School Governance: A Historical Study of the United States and Germany.

EDITED VOLUMES

Josiah Strong, *Our Country,* John Harvard Library.
With Genovesi, Bjorg B. Gundem, Manfred Heinemann, Torstein Harbo, and Tonnes Sirevag, *History of Elementary School Teaching and Curriculum,* International Series for the History of Education, vol. 1.
With Fritz-Peter Hager, Marc Depaepe, Manfred Heinemann, and Roy Love, *Aspects of Antiquity in the History of Education,* International Series for the History of Education, vol. 3.
With Henry Geitz and Jürgen Heideking, *German Influences on Education in the United States to 1917.*
With Jürgen Heideking and Marc Depaepe, *Mutual Influences on Education: Germany and the United States in the Twentieth Century,* Special Issue of *Paedagogica Historica,* vol. 33.

INDEX

References to illustrations and tables are printed in **bold**